The publisher gratefully acknowledges the
generous support of the Ahmanson Foundation
Humanities Endowment Fund of the University
of California Press Foundation.

Pineapple Culture

THE CALIFORNIA WORLD HISTORY LIBRARY

Edited by Edmund Burke III, Kenneth Pomeranz, and Patricia Seed

Pineapple Culture

A HISTORY OF THE TROPICAL

AND TEMPERATE ZONES

GARY Y. OKIHIRO

UNIVERSITY OF CALIFORNIA PRESS

Berkeley   Los Angeles   London

University of California Press, one of the most distinguished
university presses in the United States, enriches lives around the world
by advancing scholarship in the humanities, social sciences, and natural
sciences. Its activities are supported by the UC Press Foundation and
by philanthropic contributions from individuals and institutions.
For more information, visit www.ucpress.edu.

University of California Press
Berkeley and Los Angeles, California

University of California Press, Ltd.
London, England

Library of Congress Cataloging-in-Publication Data

Okihiro, Gary Y., 1945–.
Pineapple culture : a history of the tropical and temperate
zones / Gary Y. Okihiro.
p.   cm.—(The California world history library ; 10)
Includes bibliographical references and index.
ISBN 978-0-520-25513-5 (cloth : alk. paper)
1. North and south.   2. Imperialism—History.   3. Space and time—
History.   4. Culture conflict—History.   5. Tropics—History.
6. Pineapple—History.   7. Hawaii—History.   8. Hawaii—Relations—
United States   9. United States—Relations—Hawaii.   I. Title.
CB261.O38   2009
909'.093—dc22   2008043139

Manufactured in the United States of America

18   17   16   15   14   13   12   11   10   09
10   9   8   7   6   5   4   3   2   1

The paper used in this publication meets the minimum requirements
of ANSI/NISO Z39.48–1992 (R1997) *(Permanence of Paper)*.

For My Family,

Three Generations of Workers in Hawaiian Pine:

Kame Kakazu

Kashin Kakazu

Alice Shizue Okihiro

Ellen Kiyoko Nitta

Edward S. Kakazu

Joyce Ayako Kakazu Villegas

Gary Y. Okihiro

Faith Okihiro Lebb

Karen N. Oshiro

Stephen R. Oshiro

Alan K. Oshiro

CONTENTS

ILLUSTRATIONS

FIGURES

MAPS

ACKNOWLEDGMENTS

I remain grateful to Niels Hooper, whose belief in this project, a planned trilogy, has ensured its home with the remarkable University of California Press. Additionally, his colleagues at the press, notably Suzanne Knott and Rachel Lockman, have been exceptionally efficient and helpful. I am also grateful to John Thomas for his astute copyediting. A subvention from Ken Hakuta contributed to this book's illustrations and maps. DeSoto Brown of Bishop Museum (Honolulu), Amy Hau of the Noguchi Museum (Long Island City, New York), and Dore Minatodani and Joan Hori of Special Collections, Hamilton Library, University of Hawai'i (Mānoa), provided me with vital assistance and information from their respective collections. I should also mention in gratitude the staff of Columbia's libraries, especially those of the Rare Book and Manuscript Library and Avery Library, and archivists at the Hawai'i State Archives, Bishop Museum, and University of Hawai'i (Mānoa). Janet L. Appel, director of the Shirley Plantation (Charles City, Virginia), Dawn Bonner of the Mount Vernon Ladies' Association (Mount Vernon, Virginia), and Liesel Nowak of the Monticello/Thomas Jefferson Foundation (Charlottesville, Virginia) helped me with pineapple artifacts and motifs at their sites. Jeremiah Trinidad-Christensen of Columbia's Lehman Library produced splendid maps to my specifications, for which I am extremely grateful. Leah Sicat, who worked as my research assistant for a summer, was meticulous in paging through the *Ladies' Home Journal* from June 1933 to December 1945, and my longtime graduate assistant, Elda Tsou, despite having completed her degree and secured a real job, continued to guide me through the thicket of permissions, releases, and consents. My colleagues Jean J. Kim (Dartmouth) and Samuel K. Roberts (Columbia) helped steer me in the sea of tropical medicine, and my colleagues at Cornell, Viranjini Munasinghe and Vladimir Micic, and their student assistants, Ming Mai Jiang and Liana Rose Chin, were generous and helpful; Enna G. Henriquez read and commented upon a draft chapter; Colin Isamu Ritch Okihiro accompanied me on my photographic searches on Oʻahu; and countless discussions with Marina A. Henriquez about my writings generally and this work in particular have sharpened my resolve and prose. Finally, I am cheered daily by the pineapple, papaya, hibiscus, plumeria, anthurium, and ki—remigrants from Costa Rica and Hawaiʻi—that thrive in our New York City apartments, implanting the diasporic tropics in the alternately cold and hot temperate zone.

Introduction

Pineapple Culture follows *Island World,* the first volume of my trilogy on history and its conventions of space and time. As I note in the introduction to *Island World,* history's discipline involves "a linear progression of time, frequently standing in for causality or explanation, a discrete, managed space, sometimes presumed to be unique or exceptional, and humans as subjects with volition without regard for the agencies of other life forms both within and without humanity's orbit." In my version of the past, conceived of as "historical formations," I subscribe to the understood though rarely observed precept that space and time are human creations and experiences and as such are subject to contestations and reconstructions, movements and changes, varieties and standpoints. Historical formations can provide novel angles on glib assumptions of solid space and inexorable time. History's narration, too, might benefit from interruption of its conversations with itself and infatuation for elegant objectivity—science—and retreat from overheated essentialisms—politics.

Island World examines the spatial categories of islands and continents through the engagements of Hawai'i with the United States, whereas *Pineapple Culture* interrogates the tropical and temperate zones through the discursive and material career of the pineapple.[1] The ancient Greeks, I recount, believed in a "world island" of Europe, Asia, and Africa encircled by unruly and monstrous waters, and they theorized polarities of extreme heat and cold and a middling and hence moderating temperate band. A site's airs and waters, from that perspective, determined the natures or constitutions of plants, animals, and humans. Tropical lands, "rich, soft, and well-watered," Hippocrates held, nurtured people, also gendered as women, who were "fleshy, ill-articulated, moist, lazy, generally cowardly in character," whereas the temperate zone, "bare, waterless, rough, oppressed by winter's storms and burnt by the sun," produced "men who were hard, lean, well-articulated, well-braced" and "energetic, vigilant, stubborn and independent in character." Those racialized and gendered attributions validated rankings of certain men mandated to rule and others fated to subjection and slavery, according to Aristotle. Those contentions promulgated a geography, history, and science of empire, such as the expedition of the philosopher's student Alexander the Great, whose warriors and chroniclers carved out and occupied space as far east as India.

Propelled by desires, those imperial projects assumed the power to name, classify, and rule over alien lands and peoples and their resources, and those exercises commonly imposed specious, fantastic distinctions despite the realities of observation and reflection in the light of actual contact and interaction. The mappings of foreign, unearthly, and impure climes and their inhabitants were often self's projections of the familiar and strange, the secure and the threatening. Imperial geography and history similarly functioned to inflict order upon chaos, like Christopher Columbus's imaginary "Indians" and their feminized, fertile, and recumbent "New World" that begged European men's possession, ravishment, and productivity. That trope and tropical hermeneutics were perhaps most influentially advanced by one of the admiral's admirers, Alexander von Humboldt, for whom the pleasures of the "luxuriant regions of the torrid zone" held irresistible attractions. He, in the course of his invention of the tropics, forwarded empires of industry, commerce, and botany, the latter extending from experimental and display gardens in the temperate core to collectors and collection stations in the tropical fringes. Those imperial estates converted ideas into structures and practices with real economic, political, and social consequences.

Tropical products, including sugar and fruits, were systematically planted in the tropics and harvested and conveyed to the temperate zone, along with Indian and later African and Asian specimens and workers, in the course of empire. Unintended but nonetheless isolated and conquered, like their native carriers, were the diseases and infirmities of the tropics that invaded ill-suited white bodies in the field and in the homeland. Those residues of empire posed perils to "the blood" and race, but they were the necessary risks of enormous profits and national and transnational identities, prestige, and powers. Anchoring that world-system of labor and goods were tropical plantations, commercial outposts of the empire of plants, which included prominently the pineapple. The princess of fruits and a sign of conspicuous consumption and wealth, the pineapple spread from America's interior to its coasts and north to the islands of the Caribbean and European continent with its gardens and hothouses, which mimicked the tropics in the temperate zone, and finally back to plantations in the tropical band.

Reborn in its transmission, the pineapple mirrored the conversion of its handlers, the empire builders. Prominent were the missionaries from New England and their progeny in the islands of Hawai'i, where pineapple transplants came to dominate the U.S. and world markets. Their manipulations resulted in Hawaiian dispossession and loss of sovereignty and land and the installation of a white oligarchy and its earthly kingdom of capitalism. Hawaiians, especially the commoners, challenged the alienation and theft, which was effectuated by the Republic and later Territory of Hawai'i, headed by Sanford B. Dole, a son of missionaries. And without contradiction or apology, the son of Dole's cousin and another of the "mission boys," James D. Dole, from government land set aside for white settlers, launched a company and empire of pineapples that, by affecting modernity, infiltrated the markets and homes of America's heartland from the nation's Hawaiian colony.

Dole's Hawaiian Pineapple Company, with its clean, efficient machines and cannery and its advertising campaign and distribution system, capitalized upon an image of Hawai'i and created a market where none existed before. Sanitary, nutritious, and versatile, processed and fresh pineapples promised a taste of the tropics, which, like the pineapple, had been tamed through conquest and civilization. Sold as "Hawaiian," a label assumed by white usurpers of the kingdom for legitimacy, the pineapple tendered the comforts of sun-kissed lands, soft ocean breezes, nature's abundance, sensuality, and the sweet scent of paradise. Thoroughly modern, the "Hawaiian" pineapple represented an escape from the alienations of modernity. Slighted in that image making of the pineapple as a symbol of wealth and standing, hospitality, and modernity is the fruit's materiality as an object of plunder, possession, and prestige and as an extractive crop grown on conquered soil on plantations with cheap, often imported labor. Purged portraits can lose their grit and color.

Pineapple culture embraces the totality of those endowments and movements, as a native plant, food, and medicine, a captive of the New World by the Old, a symbol and object of desire, and a product mass produced, circulated, and consumed. Positioned in this history, pineapple culture invalidates the imagined, self-serving antipodes of frigid and torrid regions devoid of humanity and the temperate, civilized band, and over time also those of civilization/Christianity and barbarism/paganism, white/man and colored/woman, and colonizer and colonized. For those oppositions, reciprocal though not always equal, like the balance of trade weighed by export and import values, are the creations and impositions of the self over the other and as such reveal the locations and relations of power. Those polarities vanish in the act of movements and engagements, including empires and their resistances, which implicate imperial geography and history and their creations, accumulations, and circuits of knowledge, capital, labor, goods, and culture.

Mapping Desires

Historically, human wants have stimulated movement. Desires have fueled travel to near and distant seas and lands in the form of exploration, trade, and conquest. The Mediterranean provides an example of such travels across space and time. The seafaring peoples, including the Greeks and Phoenicians, ventured over land and water to establish trade routes for procuring valued objects such as tin and amber, which were rare in the Mediterranean basin. Once traversed, those highways and their destinations required description to enable continued access. Those markings were, for Europeans, the birth of geography, or the naming and locating of space and its natures.

In addition to spreading commerce, the fourth-century B.C. military expedition and conquests of Alexander the Great extended the knowledge and boundaries of the Mediterranean world from the Greek homeland to the Indian extremity. "Not only were the foundations of the science [geography] laid by the Greeks," a noted geographer observed, "but it was mainly through them that the observations made in the course of military campaigns, and the knowledge gained through the spread of trade, were recorded."[1]

AIRS, WATERS, SITES

Alexander's engagements with novel lands and peoples contrasts sharply with the theoretical and philosophical knowledge favored by some ancient Greeks. About 600 B.C., philosophers in Ionia sought release from the bonds of superstition and religion through formal and systematic explanation. Assuming that *kosmos,* or order, prevailed in the universe and that there was an ultimate "nature" or substance, they re-lied upon theory and inquiry to make sense of the world. Ionian observations of the movement of the sun and planets, eclipses of the sun and moon, and the fixed and changing locations of stars led to such ideas as the zodiac and the influence of heav-enly bodies over earthly affairs, the disk (flat) shape of the earth, and time (sundial and seasons). Anaximander drew the first map of the earth on the basis of that learn-ing, and Hecataeus later revised that "world island," thereby establishing the science of mapmaking.[2]

Centered on a disk girdled by an ocean stream, the island was inhabited by Scythi-ans to the north and Ethiopians to the south, Indians in the east, and Celts to the west. Beyond the habitable lands was the uninhabitable north, where extreme cold made life impossible, and the uninhabitable south, where inordinate heat prevailed. Greece, situated between the excesses of hot and cold, was ideally tempered and thus favored.

Pythagoras, an Ionian who migrated to Italy in about 530 B.C., deduced through abstract reasoning that the earth was a globe and not a flat disk, and applying the Greek division of the heavens into zones he delineated an equatorial, or summer, zone that was uninhabitable due to the heat, a polar, or winter, zone that was uninhabitable be-cause of the cold, and an inhabitable temperate zone between those extremities.[3] Ex-panding upon that notion of climatic zones and life's possibilities, Hippocrates, in his *On Airs, Waters, and Sites* (c. 410 B.C.), discussed the effects of climate on human health and character (nature). A blend of theory with observation, Hippocrates' remarkable treatise and explanation held that climate shaped the physical and biological world, including the physiques and natures of people. To illustrate that determining power of climate, Hippocrates contrasted Europe with Asia. The mild and uniform climate of Asia, he noted, with its hot and stagnant air and water, nurtured lush vegetation but laziness among the people, who appeared yellow as if suffering from jaundice. "With regard to the lack of spirit and courage among the inhabitants, the chief rea-son why Asiatics are less warlike and more gentle in character than Europeans is the uniformity of the seasons," the "father of medicine" explained. "For [climatic] uni-formity engenders slackness, while variation fosters endurance in both body and soul; rest and slackness are food for cowardice, endurance and exertion for bravery." Fur-

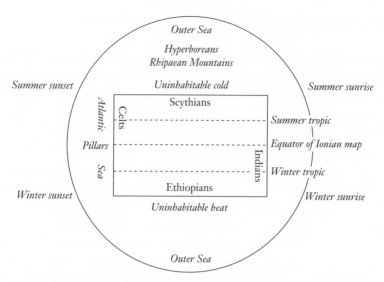

MAP 1. The Ionian "world island" with its possibilities and limits as conceived by Anaximander, c. 580 B.C. Adapted from J. Oliver Thomson, *History of Ancient Geography* (Cambridge: Cambridge University Press, 1948), 97.

thermore, their form of government, which is despotism, mirrors and reinforces that "slackness." "Courage, endurance, industry and high spirit could not arise in such conditions," Hippocrates concluded, "but pleasure must be supreme."[4]

Europeans, by contrast, experience frequent and sharp seasonal changes, which in turn favor "the greatest diversity in physique, in character, and in constitution." In Asia, where the land is "rich, soft, and well-watered," the people are "fleshy, ill-articulated, moist, lazy, generally cowardly in character." But in Europe, where the land is "bare, waterless, rough, oppressed by winter's storms and burnt by the sun," there you will find "men who are hard, lean, well-articulated, well-braced, and hairy; such natures will be found energetic, vigilant, stubborn and independent in character and in temper, wild rather than tame, of more than average sharpness and intelligence in the arts, and in war of more than average courage." "Take these observations as a standard when drawing all other conclusions," Hippocrates advised and promised his readers, "and you will make no mistake."[5]

By the time of Aristotle, those ideas of the island world were a commonplace, and although he devoted little attention to geography Aristotle considered the earth in relation to other heavenly bodies in "Meteorologica" and "On the Heavens." Like Pythagoras he believed in a spherical earth located at the center of the universe, and like Parmenides he held that there were a torrid and two frigid zones where life was impossible, and between them temperate zones north and south of the equator.[6] Aristotle concurred with and enlarged upon Hippocrates' contention that climate molded human nature and institutions in his disquisition on politics. The cold climate to the north of Greece, the philosopher taught, bred Europeans, who are "full of spirit, but wanting in intelligence and skill; and therefore they retain comparative freedom, but have no political organization, and are incapable of ruling over others." Whereas to the east, the uniform climate gave birth to Asians, who are "intelligent and inventive, but they are wanting in spirit, and therefore they are always in a state of subjection and slavery." In fact, Aristotle maintained, Asians are "by nature slaves," and they "do not rebel against a despotic government." The Greek "race," on the other hand, situated between Europeans and Asians, "is likewise intermediate in character, being high-spirited and also intelligent. Hence it continues free, and is the best-governed of any nation, and, if it could be formed into one state, would be able to rule the world."[7]

IMPERIAL SCIENCE

The empire envisioned by Aristotle was being created during his lifetime by one of his students. Alexander the Great crossed the Hellespont in 334 B.C. with an army of 40,000 men and, for the next eleven years, advanced across Asia. A voyage of exploration and conquest, that expedition included soldiers but also scholars and scientists who recorded the conqueror's exploits and the geography of the vanquished lands and peoples. Although a Macedonian, not a Greek, Alexander and the kings of Macedon claimed Greek lineage, and his war against Persia was ostensibly waged to avenge the sufferings inflicted upon the Greeks during the Persian wars about a century and a half earlier.[8] Turned back from reaching the Ganges only by the threat of mutiny among his own ranks, Alexander fell ill and died in Babylon in 323, leaving behind a divided empire of oft-warring factions and kingdoms because of dissension among his generals.

The accumulated observations of plants, animals, lands, and peoples from Alexander's march enabled a larger and more detailed apprehension of the world. Uncertainties of distances and latitudes from Alexander's expedition, however, resulted in erroneous mappings, such as the one drawn by Eratosthenes (c. 276–196 B.C.), the "father of systematic geography" and head of Alexandria's immense library. Relying on the centuries-old causal connection between climate and character, Eratosthenes applied the descriptions of Alexander's chroniclers to his mapmaking and placed Ethiopia and India on the same latitude because, according to the archive of accumulated knowledge, they shared a climate, plants and animals, and black people.[9] The result reveals the Greek reliance on theory to help organize and make systematic observation and the persistence and power of certain ideas, such as climatic and geographic determinism.

Despite the Roman Empire's spread, Roman understanding of the world remained largely derivative of Greek learning, including ideas of climatic zones and their life forms. The ancient Romans conducted a flourishing trade with India, and reports like the remarkable *Periplus of the Erythrean Sea* (Circuit of the Indian Ocean, c. A.D. 50), written by an anonymous Roman subject from Egypt, gave detailed descriptions of the Indian Ocean trade, its ports, products, and peoples.[10] Still, wrote one historian, "factual knowledge was limited, and it was frequently impossible for the most sincere and critical commentators to separate myth from fact, or to distinguish clearly one Eastern area or people from another. India, as distinguished from China, was the scene of marvels and the habitat of monstrous animals and peoples." And, as if to prod the imagination further into the realm of the fantastic, Romans imported parrots, monkeys, elephants, rhinoceroses, furs, skins, rugs, ivory, pearls, pepper, cinnamon, and possibly bananas, which may have been grown in Rome. During the Roman age, accordingly, "the myth of Asia as a land of griffons, monsters, and demons, lying somewhere beyond the terrestrial Paradise, slowly enmeshed the popular imagination of medieval Europe and gradually penetrated the popular literature of the crusading era."[11]

NATURAL LAW

After the breakup of the Roman Empire and up to the sixteenth century, the Christian church and its theology mediated geographic thinking, distancing itself from Ionian and Athenian science and approaching a return to faith.[12] That spiritual kingdom, although powerful and long-lived, failed to dominate absolutely. Legal scholar Jean Bodin "moved away from the authoritarian [religious] toward the natural [secular]," according to his translator, and although he believed that God moved the heavenly bodies, he also distinguished between knowledge derived from the divine and the human, or "natural." Bodin's *Method for the Easy Comprehension of History,* first published in 1566, strived to uncover "universal law," which he believed was revealed in the recounting of human experience. Bodin, a Renaissance man, marked a transition from the medieval (faith) to the modern (science), and he saw geography as the gateway for an "easy comprehension" of history.[13] The major geographic tenet and "universal law" subscribed to by Bodin was the climate's sway over human affairs and constitutions drawn from Hippocrates and his intellectual descendants.[14]

Bodin modified the ancient Greek notion of habitable latitudes by dividing the earth into a sparsely populated polar zone, a thickly settled and developed temperate zone, and a tropical band. Each zone, he maintained, had its own place assigned to it by geography and the environment, and, as in Plato's republic, the whole functioned to create universal harmony, unity amidst diversity in a "republic of the world."[15] The center of his analysis, however, and the latitude of superior peoples was the temperate zone. With the European "discovery" of America, Bodin's world extended beyond the Celts to the west, and he knew from voyages that the tropics and equatorial zones, despite "incredible heat," were habitable.[16] Still, like Eratosthenes, Bodin held erroneously that Ethiopians and Indians lived in the same latitude and were consequently black by "seed," and that men in cooler climates possessed "inner warmth," which ignited energy and enabled robust activity, while Africans were devoid of "internal heat" and were thus lazy though bearing "work and heat patiently."[17]

Bodin represented a curious mixture of traditions of science and religion when he declared that climate and celestial bodies failed to exercise complete control over humans, "yet men are so much influenced by them that they can not overcome the law of nature except through divine aid or their own continued self-discipline." Further, he believed that the environment shapes human physiques, body forms reveal "habits of the mind," and those characteristics constitute "the inborn nature of each race." Reduced to its essence, body type, intellect, and natures are "in the blood" and "from blood." "The elements are disturbed by the power of the celestial bodies," Bodin wrote in summation, "while the human body is encompassed in the elements, the blood in the body, the spirit in the blood, the soul in the spirit, the mind in the soul."[18]

NATURAL INEQUALITY

Bodin, in his search for universal law, provided, in the words of a recent study of the modern idea of race, "a history of mankind divided up into peoples and dispositions arranged according to astrological and astronomical influences, climate, language, geographical location."[19] Those groupings, loosely marked by "the inborn nature of each race," would later achieve solidity as "races" by the scientific classifiers of the eighteenth century. In 1775, Johann Friedrich Blumenbach published his doctoral dissertation, *On the Natural Variety of Mankind,* for which he is considered the father of anthropology and, more specifically, physical anthropology. In that edition, Blumenbach described four "varieties" of humans—European, Asian, African, and American—and the factors that led to that diversity—climate, mode of life, and hybridity. Climate, the anatomist claimed, can alter body shape and therewith culture, though humans and animals can move to unfamiliar climatic zones and change their physiques. Likewise, mode of life and hybridity can produce new body forms. Differentiation, however, has its limits, Blumenbach cautioned, and, despite "the unity of the human species," there are four "mere varieties" differentiated by "the structure of the human body."[20]

Europeans, Asians, Africans, and Americans, Blumenbach noted, differ in "bodily constitution, stature, and colour," and those features are due "almost entirely to climate alone." Thus, thinking like Hippocrates, "in hot countries bodies become drier and heavier; in cold and wet ones softer, more full of juice and spongy," and that is why the bones of Ethiopians are "thick, compact, and hard." Where the climate is mild (as in Asia), people are smaller and less fierce, and Ethiopians have black skin color because of the climate, temperature, soil, and mode of life. That idea, Blumenbach acknowledged his debt and pedigree, "is the old opinion of Aristotle, Alexander, Strabo, and others." But Ethiopians can lose their blackness by moving to northern Europe, where their skin will become brown, and white Spaniards living in the "torrid zone" have "degenerated" to the color of the soil. These examples point to changes induced by the environment but not the creation of new species.[21]

Blumenbach's third edition, issued in 1795, proposed five principal "varieties" of mankind—Caucasian, Mongolian, Ethiopian, American, and Malay. And it highlighted a concept first introduced in the earlier version, the notion of "degeneration" as seen in skin color, hair texture, stature, bodily proportion, and skull shape, caused by diet, mode of life, hybridity, and above all, climate, which was "almost infinite" in its power over bodies. Climate has "the greatest and most permanent influence over national colour," Blumenbach stated, and skin color is the most reliable indicator of "variety" because, "although it sometimes deceives, [skin color] still is a much more constant character, and more generally transmitted than the others." Europeans are white, thus, Mongolians yellow or "olive-tinge," Americans copper or "dark orange," Malays tawny, and Ethiopians tawny-black to "pitchy blackness." In the "torrid zone," abundant heat and carbon induce the liver to produce an excess of bile, or "black bile" in Bodin's words. Consequently, "the temperament of most inhabitants of tropical countries is choleric and prone to anger."[22]

Unlike Bodin, whose historical schema of progress and stagnation relied upon an almost mystical quality of blood, Blumenbach catalogued differences on the basis of his measurements of skulls and cranial capacities, teeth, breasts, penises, hands, feet, and statures.[23] The result was a scientific nomenclature and classification of human types, which provided a foundation for the studies that followed. And by citing "degeneration" as the process involved in differentiation, Blumenbach established a hierarchy among his five varieties of humankind. As he explained, his use of the word "Caucasian" to denote his former "European" derived from the Caucasus Mountains, because that area produced "the most beautiful race of men, I mean the Georgian."[24] All others represented "degenerations" from that ideal type, like the Ethiopian of Bodin's description, who was "very keen and lustful" and "small, curly-haired, black, flat-nosed, blubber-lipped, and bald, with white teeth and black eyes."[25] And Blumenbach's Cau-casian hair was soft, long, and undulating, whereas Mongolian and American hair was black, stiff, straight, and scanty.[26] Furthermore, despite a recognition of the essential unity of humans as a single species and the ties between "varieties" that "run together," Blumenbach offered a history, an evolution of human types distinguished by physical and behavioral characteristics that were transmitted through reproduction and formed correspondences with mappings of climates and constitutions.[27]

Taking up Blumenbach's theory of degeneration and Bodin's of "blood," Arthur de Gobineau wrote as early as 1853 *The Inequality of Human Races*. His principal concern, having witnessed the global spread of Europeans and their implantations in the tropical band, including trade outposts, colonies, and interracial encounters, was the "inequality" of the races and the lowering of white blood by inferior elements. "The pure-blooded yellow and black races," "backward" and "weaker" strains, have spread from the tropics to the temperate zone to all corners of the globe, and because their labor is indispensable to their "masters" they coexist and "the mixture of blood finally takes place." As a result, the "primordial race-unit is so broken up and swamped by the influx of foreign elements" that "the people has no longer the same intrinsic value as it had before, because it has no longer the same blood in its veins, continual adul-terations having gradually affected the quality of that blood." Like nations that have died because of degeneration, whites faced extinction through the impurities, which eroded their inner constitutions.[28]

Gobineau and many of his contemporaries were more impressed by the power of "blood" than that of climate or institutions, which played roles in but failed to account for "racial inequality."[29] In his scheme, society and geography formed nondetermin-ing environments in and to which humans operated and reacted, explaining how "inferior" peoples like the Hawaiians could mimic but not fully absorb civilization, and North American Indians could live in a temperate and resource-rich land for hundreds of years but fail to advance to a high state, as he defined it, of civilization.[30] Gobineau, thus, parted company with the geographic determinism of his forebears and proposed just three races based upon skin color—white, yellow, and black—with all others derivatives of those main groups. He believed in the unity of all humans, like Blumenbach, and placed white at the apex and yellow and black as "degenerations" of that ideal. "I . . . have no hesitation in regarding the white race as superior to all others in beauty," Gobineau stated categorically. "Thus the human groups are unequal in beauty [and intellect]; and this inequality is rational, logical, permanent, and indestructible." Accordingly, he claimed, there was "irreconcilable antagonism between different races," and "innate repulsion" supplied the main motive force of history, determining its course.[31]

RACE/GENDER ISOTHERMS

Whereas race mixture may have posed a problem for Gobineau, hybridization was a possible solution to the "problem of tropical colonization," for Ellen Churchill Semple, given the "debilitating effects of heat and humidity" and tropical diseases, which made the tropics the "white man's grave."[32] Semple, an interpreter of eminent naturalist and human geographer Friedrich Rätzel in her *Influences of Geographic Environment* (1911), sought to reinsert "geographic factors and influences" in the shaping of society and its institutions and to connect them with the formation of races and "ethnic stocks."[33] Geography's influence over history had fallen into disrepute, Semple regretted, because of caricatures of the science that simplified complex explanations.

Unstated in Semple's reprise of geography's primacy was the prevailing enthusiasm for race, or "blood," as in Gobineau, and the ideas known as social Darwinism and the new science of "eugenics" introduced by Darwin's cousin Francis Galton in 1883. Eugenics sought to establish through mathematics the hypothesis that bodily conformation or physique and mental abilities or intelligence were passed through the blood from one generation to another according to the Darwinian laws of natural selection. It took a social activist turn in its attempt, through laws, institutions, and behavior, to regulate and cultivate "good" gene pools and eliminate or segregate "bad" ones.[34]

"Man is a product of the earth's surface," began Semple, in that it feeds, nurtures, and molds his body, circumscribes his ideas and ambitions, and even shapes his religion, despite man's claims to have conquered nature. Rather, the environment has quietly and persistently made the man. And because all of human activity, history, takes place on the earth, historical development is "more or less molded by its geographic setting." History, in fact, is simply "a succession of geographical factors embodied in events."[35] Those geographic influences include the direct physical effects of the environment in forming diverse "races and peoples"; their size, skin color and thickness, and hair; the psychical impact of the climate on the "temperament" and character of the various races; the earth's provisions that advance or retard wealth and hence cultural and political possibilities; and the effects on human migration. Accordingly, racial and social differentiation arises from "modifications in response to various habitats in long periods of time" and the processes of natural selection and inheritance.[36]

Movements have led to separation, isolation, and differentiation but also to race mixture, assimilation, and hybridism. And despite constant migrations and "an endless mingling of races and cultures," a general pattern prevails wherein whites remain in the temperate zone and peoples of color in the tropics, although white, yellow, and red can be found in every zone, while black, mainly in the tropics. That global distribution of "races and cultures" reflected Semple's time and problem. The late nineteenth century into the early twentieth was a period of European expansion to and colonization of the tropics, stimulated in large part by the desire for commodities, markets, and labor. Semple racialized and gendered that imperial spread, white and manly, as indicative of a vigorous and strong race bent upon progress and civilization, and she framed it as a principle of geography and social Darwinism: expansion "is an expression of the law that for peoples and races the struggle for existence is at bottom a struggle for space."[37] Conversely, small, weak, and primitive races, colored and womanly, occupy limited territories.[38]

Although humans can escape "the full tyranny of climatic control" and its effects can often be overstated, Semple admitted, climate, meaning temperature and moisture, is not merely the context for people's activities but shapes their bodies, physically and psychologically. It influences their immune system and resistance to diseases, "their temperament, their energy, their capacity for sustained or . . . intermittent effort," and thus "their efficiency as economic and political agents." Man can make himself at home in any zone, but "zonal locations" (latitudes of temperature and rainfall) fix the borders of human habitation and determine "race temperament" and civilization.[39] Those ideas are a twentieth-century rendition of some of the founding geographic formulations of the ancient Greeks.

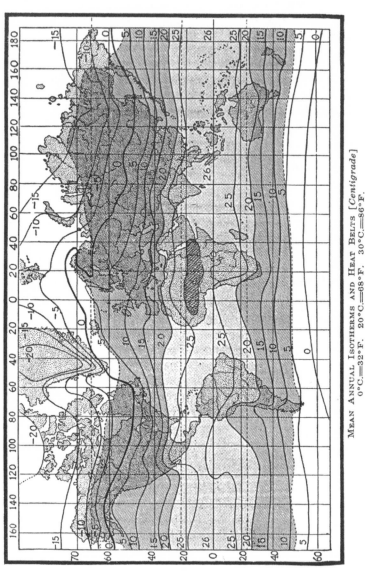

MEAN ANNUAL ISOTHERMS AND HEAT BELTS [*Centigrade*]
0°C.=32°F.   20°C.=68°F.   30°C.=86°F.

MAP 2. Like the mappings of climatic and biotic zones by the ancient Greeks, Semple's isotherms and heat belts delineate civilizations and "race temperaments." From Ellen Churchill Semple, *Influences of Geographic Environment* (New York: Holt, Rinehart and Winston, 1911), 612.

Semple's zonal locations produce and explain racializations. "The northern peoples of Europe," Semple wrote, like Hippocrates and Aristotle, "are energetic, provident, serious, thoughtful rather than emotional, cautious rather than impulsive." By contrast, and unlike the ethnocentric Greek philosophers' notions, "the southerners of the sub-tropical Mediterranean basin are easy-going, improvident except under pressing necessity, gay, emotional, imaginative, all qualities which among the negroes of the equatorial belt degenerate into grave racial faults." In the tropics, the heat "tends to relax the mental and moral fiber, induces indolence, self-indulgences and various excesses which lower the physical tone." The enervating temperature makes natives lazy, and even "energetic" whites there are drawn down the path of economic and social "retardation."[40]

"These broad belts," then, "each with its characteristic climatic conditions and appropriate civilization, form so many girdles of culture around the earth," Semple posited. The temperate band "is the seat of the most important, most steadily progressive civilizations, and the source of all the cultural stimuli which have given an upward start to civilization in other zones during the past three centuries." In the tropics, where man was born "in his primitive, pre-civilized state, he lived in a moist, warm, uniform climate which supplied abundantly his simple wants, put no strain on his feeble intellect and will." Like a womb and prison, this "nursery has kept him a child." "As the Tropics have been the cradle of humanity, the Temperate Zone has been the cradle and school of civilization," Semple summarized. "Here Nature has given much by withholding much. Here man found his birthright, the privilege of the struggle."[41]

In an age of European imperialism, Semple wrote, "nature has fixed the mutual destiny of [the] tropical and temperate zones . . . as complementary trade regions," and that empire's creation, "the privilege of the struggle," involved "the conquering white race of the Temperate Zone," whose desire for tropical products has driven the energetic race to the "productive but undeveloped Tropics."[42] According to this historical geography, then, "nature" (and in other renditions "destiny") has preordained the expansion of "the conquering white race" as if impelled by science and the laws of the natural world. Accordingly, Semple likened the "great historical movements in the form of migration, conquest, colonization, and commerce" to "convection currents" that "seek to equalize the differences and reach an equilibrium."[43] The direction of those drafts, we know from the physical world, is from (white and manly) areas of high densities, pressures, and activities to (colored and womanly) vacant and inert spaces.

Whereas Semple may not have worried over the capacity of whites to flourish in the tropical hinterland or fear the swamping of the white by migrating nonwhite races in the temperate homeland,[44] Yale geographer and president of the board of the American Eugenics Society Ellsworth Huntington shared the nineteenth-century European anxieties over race mixture and degeneration as a consequence of human movements and empire. In his *Civilization and Climate* (1915), Huntington cited race and "racial inheritance," social institutions, and, like Aristotle and others, climate in the rise of civilization. In brief, "good stock, proper cultivation, and favorable climatic conditions" produce "the fruit known as civilization." Of the three, Huntington regretted, the significance of climate as an explanation has been eclipsed by race and institutions because, this proponent of the "new science of geography" declared, of ignorance of the latest findings in archaeology, which reveal the intimate connection between climate and civilization. The clear lesson of antiquity, Huntington reported, is that "a certain peculiar type of climate prevails wherever civilization is high," as in ancient Egypt and Greece, where the climate filled the people with "a virile energy."[45]

Race, however, should not be discounted altogether, the Yale researcher cautioned, like Gobineau, for a favorable climate will not cause "a stupid and degenerate race to rise to a high level." Studies have shown that "the brain of the white man is more complex than that of his black brother," and that no amount of training can compensate for that "ineradicable racial difference in mentality." The Hampton Institute,[46] for instance, demonstrates how "the Christian spirit" and "proper training" can help but ultimately fail to overcome "the handicap of race." Analogous to the earth's diverse trees and fruits, those racial differences constitute complementary parts to the whole. "Initiative, inventiveness, versatility, and the power of leadership," Huntington declared, "are the qualities which give flavor to the Teutonic race. Good humor, patience, loyalty, and the power of self-sacrifice give flavor to the negro."[47] Samuel Chapman Armstrong, Hampton's master teacher and commander of African American troops during the Civil War, similarly believed that "the Negro," loyal

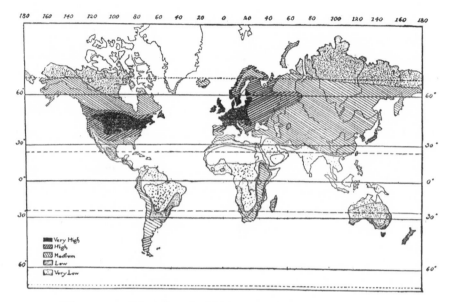

MAP 3. "Human energy" flows from the "climatic energy" concentrated, as mapped by the ancient Greeks on down, in temperate latitudes. From Ellsworth Huntington, *Civilization and Climate* (New Haven, Conn.: Yale University Press, 1915), 142.

and self-sacrificing, made good soldiers: "They are noble under leadership, often wonderful in emergencies."[48]

Climate, nonetheless, exerts itself over human affairs regardless of race, affecting both whites and blacks, and in fact it "controls the phenomena of life from the lowest activities of protoplasm to the highest activities of the human intellect." When shouldering the "burden" of his race, gender, and age—the expansion of whites from the temperate to the tropical girdle because of the region's enormous wealth—Huntington expressed two concerns: Tropical "natives," he noted, are "dull in thought and slow in action," and "experience shows that the presence of an inferior race in large numbers tends constantly to lower the standards of the dominant race."[49] Accordingly and contrary to Semple's proposal, interracial "breeding" is not a eugenic solution to that intercourse. Further, the hot climate induces "tropical inertia" on white minds and bodies, lowering their intellectual capacities and physical energies, their sexual and moral inhibitions, and their resistance to tropical diseases. The temperature and humidity of the tropical band and its "native races" thus threaten to weaken the white stock through fatigue, dilute and pollute it through miscegenation, and sap it of "human energy," which is the engine for civilization and which, in turn, derives from "climatic energy."

Huntington closed with an apocalyptic vision of an approaching climatic change, as was revealed in the archaeological record, which might favor places like Egypt, Mesopotamia, and Guatemala and result in "a chaos far worse than that of the Dark Ages" in which "races of low mental caliber may be stimulated to most pernicious activity, while those of high capacity may not have energy to withstand their more barbarous neighbors."[50]

SPATIAL VIOLATIONS

The two major problems of the age of European empire, the late nineteenth and early twentieth centuries, involved the presence of whites in the tropics and its reciprocal, the movement of people of color from the tropical to the temperate zones. Those concerns spawned numerous studies on the subject, those by Ellen Churchill Semple and Ellsworth Huntington among many others. Premier imperialists, the British were among the leaders in this work, including historian Charles H. Pearson and sociologist Benjamin Kidd.

In agreement with Huntington's dire future for white supremacy, Pearson recalled that white desires for tropical products propelled their movement from the temperate to the tropical band, where they introduced modern industry and medicine and hygiene and thereby increased native productivity and population growth. Given those developments, Pearson predicted, "the day will come, and perhaps is not far distant, when the European observer will look round to see the globe girdled with a continuous zone of the black and yellow races, no longer too weak for aggression or under tutelage, but independent, or practically so, in government, monopolising the trade of their own regions, and circumscribing the industry of the European." Further, those "black and yellow races" will "throng" to the "salons of Paris" and "English turf," intermarry with whites, and "we shall wake to find ourselves elbowed and hustled, and perhaps even thrust aside, by peoples whom we looked down upon as servile and thought of as bound always to minister to our needs."[51]

Benjamin Kidd proposed to shift Pearson's stress on the education and expansion of tropical peoples to the "development in progress amongst the Western peoples." In his *Social Evolution* (1894) and *The Control of the Tropics* (1898), Kidd issued "a clear call of duty" interlaced with "moral force" for white management of the tropical band. Independence in Central and South America had led not to "a state of high social development," he charged, but to political corruption and economic bankruptcy, and he contrasted that "degeneration" with the advances in European natural and social sciences that had brought coherence to disparate bodies of learning, as demonstrated in the work of Charles Darwin and its applications to human societies. Altruism and the principle of "the native equality of men," Kidd assured, governed white relations with "the coloured races outside the temperate regions," and not conquest, occupation, and exploitation, which contradicted the ideals of civilization and were "anachronisms." European development of the tropics, "the richest region of the globe," was made necessary only by "a permanent state of uncertainty, lack of energy and enterprise amongst the people, and general commercial stagnation."[52] Yet the burden of white interest in the torrid zone, Kidd conceded in the end, was based upon the fact "that the complex life of the modern world rests upon the production of the tropics to an extent which is scarcely realized by the average mind."[53]

A contemporary of Pearson and Kidd, Scottish geographer James Bryce claimed that human migration was "the most potent factor in making the world of to-day different from the world of thirty centuries ago" and in shaping the future of "the race." The march of Alexander the Great through Asia and the Spanish colonization of America, he wrote, were examples of "permeation," the movement of "a certain . . . number of persons of a vigorous and masterful race into a territory inhabited by another race of less force, or perhaps on a lower level of culture."[54]

With the United States poised to join the club of European imperialists by acquiring territory in the Caribbean and Pacific, Bryce was asked to offer some advice based on the British and European experience. Climate, he replied, determines the colonial profile, which in turn corresponds with race and form of government. A temperate colony is one in which whites can "live and thrive and bring up healthy children," whereas a subtropical colony is one in which whites can maintain themselves but "cannot do hard and continuous work." In a tropical colony, whites are "forbidden by the heat not only to support open-air labor, but also to retain its original robustness of mind and body." The temperate colony, now completely occupied, he continued, is the "natural home of the European races," whereas colored people or those of "a different blood" belonging to "a lower type of civilization" are most numerous in the subtropical colony, and more marked is the race distinction in a tropical colony where a small group of "civilized" white men rule over an enormous mass of "savage or semi-civilized men . . . of a different color."[55] Although Bryce's advice on colonial acquisition and rule provoked commentary, virtually unquestioned was his claim that climate shaped the contours of the colonial project and formed correspondences with race and civilization.[56]

That critique would come later in the work of physicians and scientists who found that white bodies, although susceptible to tropical diseases, could adapt and survive in the torrid zone, contrary to the assertions of such writers as Kidd and Bryce.[57] Disease, not climate, was the enemy of whites in the tropics, a 1919 report in the *Journal of Tropical Medicine* emphatically announced.[58] Building upon that finding, J. W. Gregory, a University of Glasgow geologist, noted that climate has worked an "instinctive sorting" of racial types in which "the tropics are the natural home of the coloured races and the temperate regions that of the white races." That "racial distribution" arises from the geographic principles of climate and population differentiation and density and from the struggle for land and resources in which the inferiors survive because they can get by with less. And yet the body temperature of white American soldiers in the Philippines differs little from that of other whites in the United States, and those soldiers have shown an ability to acclimatize to the heat and humidity of the tropics. In addition, studies indicate that the lungs, kidneys, and nerves of whites undergo no important change in the tropics, as was commonly believed, and some insurance companies have concluded that the higher rates they charge for living in the tropics are unmerited.[59]

White penetration of the tropics is, then, physiologically possible, Gregory concluded, but in the cohabitation of white men and colored women rests the danger. Miscegenation, he warned, is "mischievous and dangerous," as if against nature, because "the interbreeding of widely different races of mankind produces inferior offspring." Instead, the professor advised, the white should secure "as his home" the continents of Europe, North America, and Australia, and from that base "for the benefit of all, continue to conquer the forces of Nature and thereby strengthen the broad foundations of civilization."[60]

LIFE'S COMPLEXION

French historian Lucien Febvre took up that concern of modernity—the relations between humans and nature—and favored human agency over the environmental determinism of earlier periods of European thought. The ancient Greeks, Febvre recalled, including Hippocrates, Plato, and Aristotle, fixed human bodies (races) and their societies (cultures, civilizations) to the land (environment) bounded by climatic zones or latitudes. Jean Bodin, he pointed out, represented an advance over the absolutism of both Greek tradition and Christian dogma and superstition, and Friedrich Rätzel helped to shift geography's center from the physical to the human in his "anthropogeography." "It is not true that four or five great geographic influences weigh on historic bodies with a rigid and uniform influence; but at every instant and in all phases of their existence," Febvre asserted, "through the exceedingly supple and persistent mediation of those living beings endowed with initiative, called men, isolated or in groups, there are constant, durable, manifold, and at times contradictory influences exercised by all those forces of soil, climate, vegetation—and many other forces besides—which constitute and compose a natural environment."[61]

Determinism, for Febvre, is not history; rather, "the very basis of history" is the understanding of humans as "efficacious agents" of somatic and social constitutions and changes. Additionally, although climate has produced the tropical and temperate bands with their distinctive life forms and possibilities, it does not mandate human imagination or behavior, nor is it the sole factor in a multiplicity of complex influences shaping human nature and society. Geographic categories exceed the bounds of climatic latitudes such as mountains and coasts, islands and continents with their diverse plant and animal life and human economies of hunting and fishing, pastoralism and agriculture. Boundaries, in fact, are human creations, because along the frontiers formations such as "the Eastern world, the world of Islam, the Asiatic world" and European civilization interact, blurring distinctions. The move from "determination" to "only approximations and probabilities," Febvre admitted, might lose "the beautiful simplicity and certainty of the mechanical explanations" but gain "a richer and more complex view, better matched with the exact complexion of the phenomena of life."[62]

It was precisely that intention to capture "the exact complexion of the phenomena of life" that led to the search for formal and systematic explanations freed from superstition and religion among the ancient Greeks. And although theoretical and philosophical in orientation, these explanations more closely modeled reality in actual encounters and insisted upon evidence and demonstration. Geography, as conceived in Europe, grew from that imperative to apprehend the world, its lands and waters, its resources both animate and inanimate, its peoples, and its place in the heavens. Human movement, including migration, exploration, trade, and conquest, enabled and required mappings of familiar and novel climes and peoples. That geography posited a circular and later spherical island world of lands surrounded by seas and divided by "airs, waters, and sites" into habitable and uninhabitable girdles of freezing cold, moderation, and scorching heat. The sole causal connection ancient Greeks and their intellectual descendants established between environment and human physiques, natures, and organizations, as Febvre pointed out, eventuated in the influential falsehood of race and its differentiation and hierarchy.[63]

The reciprocal of those movements from the homeland to the edges of the known world were imports, including goods—the objects of desire—and peoples, from the peripheries to the centers. Nature, as analogized by Ellen Churchill Semple to convection currents, is insufficient as an explanation for that traffic between the temperate and tropical bands, and "destiny," whether manifest or divinely ordained, fails to account for the naming and characterizing of those latitudes and their peoples. As pointed out by Lucien Febvre, those acts are complex and multifaceted and are, above all, interventions of "living beings endowed with initiative, called men." In fact, white desires for tropical products and the region's enormous array of untapped resources and fecund lands provisioned trade vessels and seeded anchorages for colonies. And from that penetration arose problems of white morbidity and the "breeding" of mixed and hence inferior stock in the tropics and the infiltration of nonwhite bodies, diseases, and cultures into the white turf and gene pool. Prodigious profits and perils were close companions, then, in this quest for tropical empires.

Empire's Tropics

Histories, like maps, name, order, and confer meanings to space. By and large, they create, occupy, and populate space with historical actors, and they narrate events, selected to be sure, as they unfold over time. As social and human inventions and enactments, they are revealing of power and its manifestations. In that way, although historians might pose as impartial scribes, their landscapes and staging of space might more accurately bear the name "imperial history," as correctly identified by essayist Paul Carter, who reserves the label for a particular variety of history that gains legitimacy through the logic of cause and effect and of creating order from chaos. The primary object of imperial history, he explains, "is not to understand or to interpret: it is to legitimate."[1] When historical writing assumes mastery over time and space, imposing its spatial boundaries, linear chronologies, and causal explanations, it warrants the title "imperial."

"There is much to be said for the view that Victorian geography was the science of empire par excellence," wrote a geographer. Begun in the 1830s during anxieties over the British Empire, the Royal Geographical Society of London staked its claim to existence as of "first importance" to "mankind in general" and in particular "paramount to the welfare of a maritime nation like Great Britain, with its numerous and extensive foreign possessions." One of its first presidents, W. R. Hamilton, expanded upon that ambition by noting that geography was "the mainspring of all the operations of war, and of all the negotiations of a state of peace." Any imperial power, he reasoned, required knowledge about the lands and waters of its colonies and their opportunities for commerce, for "enlarging her powers of civilizing yet benighted portions of the globe, and for bearing her part in forwarding and directing the destinies of mankind."[2] In that way, nineteenth-century geography was a discipline of empire, remaking the world literally and figuratively in the name of progress.[3]

EUROPE'S ASIA

Desires for the goods of Asia prompted the period of European empire—glossed as "the age of exploration," launched in the late fifteenth century—and the mapping of those routes for commerce and the recreation of the "New World" in the image of the "Old" began long before Christopher Columbus's island landfall in 1492.[4] The admiral's real gift to his royal patrons was not the thrill of discovery but the lands and labor upon which to build a New Spain and inaugurate a Spanish empire. Columbus's topography too was a flexing of imperial powers. His imaginary "Indians" and tracings of sites he knew as Asia to his death were overlays of medieval Europe's patterns of the Orient, which named, described, and assigned value to that world, establishing a "tropical hermeneutics," as astutely advanced by one geographer.[5]

Hermeneutics, or interpretation, he explained, arises from fabricated fissures between the self and other, the familiar and alien, and a tropical hermeneutics maps the tropics as an imaginary born of geographic difference and as a physical location, the band between parallels of latitude, 23°26' north (Tropic of Cancer) and south (Tropic of Capricorn) of the equator.[6] In their attempt to manage the strange from the standpoint of the familiar, Columbus's logs constituted a science of nature, which abetted and accompanied Europe's imperial projects both abroad and at home.[7]

The fabulous East, the vicinity of Paradise the admiral thought he had discovered, he likened to the biblical Garden of Eden inhabited by women and children, innocent in their nakedness. He feminized its fertile and fruitful fields: "Like a woman's nipple," as he fancied the Orinoco River—from whence, he was sure, flowed the waters of Eden found at its source.[8] And in his moment of first discovery on October 12, 1492, in the islands called the Bahamas, as a man and master, Columbus took "possession" of them.[9] That grab, he claimed, was in compliance with his commission from Spain's Catholic rulers to "discover, take possession, govern, and trade," and it was moreover an act of charity as inscribed on his coat of arms: "To Castile and León [his royal patrons] Columbus gave a new world."[10] Revealing of the nature of his instructions, to "discover," in the Spanish of the admiral's times, referenced the scouting of territory in anticipation of battle to gain strategic advantage over one's opponent.[11]

Europeans—from the ancient Greeks, as seen in the writings of Hippocrates, to the fifteenth-century empire builders led by Portugal and Spain—commonly believed in an Asia that was both rich in plant life and material resources and impoverished in human initiative, civilization, and culture. Those ideas in medieval Europe translated into the myth of Asia as the land of lush vegetation, opulence, and leisure— a terrestrial Paradise—and as a state of decline and decadence and a place of fantastic and monstrous creatures. *The Travels of Sir John Mandeville* (1356) described an Orient bloated with gold, silver, precious stones, one-eyed and headless beasts, giants, pygmies, and cannibals. *Mandeville* remained the standard text on Asia and was widely read for more than 150 years by the likes of Columbus, English "adventurers" Martin Frobisher and Walter Raleigh, and Flemish cartographer Gerhardus Mercator, who were all great admirers of the book.[12] Such hermeneutics and maps of the Orient that preceded their setting foot on Asia proper guided European "explorations" of the "New World."[13]

The thought that America was not Asia was speculated upon by Pietro Martir de Anghiera, who titled his 1493 account of Columbus's voyage *De Orbe Novo* (Of the New World), and by the admiral himself, who, during his second voyage, announced that he had reached a land unknown to the ancients.[14] That new geography was more firmly established in 1513 when Francisco Nuñez de Balboa crossed Panama's isthmus and waded into the Mar del Sur (South Sea), and six years later when Ferdinand Magellan rounded America's southern tip and entered the ocean he called "the Pacific" because of its waters' calm. Despite that fracturing of the world into discrete landmasses separated by seas and the globe's consequent broadened girth, Europe's creation of Asia continued to exert a powerful influence over its observations of and activities in America.[15]

Perhaps disappointed, Columbus admitted after having made his first landfall, "I have found no monsters," but he was hopeful of finding just beyond the Bahamas, with the help of Lucayo navigators, "people . . . born with tails." Instead of beasts, he kidnapped and removed to Spain six "Indians" as specimens from presumably India for display in the Spanish court.[16] The following year, the admiral headed an expedition to establish a permanent settlement in America and forced the natives of Hispaniola, which later became Haiti and the Dominican Republic, to forsake their food cultivation to find and dig gold for him, a demand that resulted in famine, many deaths, revolts, and harsh reprisals. And in 1496, expanding upon his earlier traffic, he conveyed to Spain 550 Tainos, the 350 survivors of that trans-Atlantic passage to serve as slaves.[17] Two years later, members of Amerigo Vespucci's first expedition returned to Spain with 222 natives for sale as slaves. The lands Vespucci plundered geographers named "America" in honor of the trafficker, and the nomenclature became a European convention from the seventeenth century onward.[18]

Europeans, prominently Spaniards, pursued America's gold and silver with a single-minded ruthlessness that befitted the dreams of wealth that underwrote the search for a route to Asia and its treasures.[19] Instead, America's precious metals, exacted from Indians, were the means by which the Spanish conducted trade in Asia, where notably the Chinese possessed little desire for European manufactures. America's peoples, when sold as slaves in Europe, helped to defray costs and justify the expenses of expeditions.[20]

ASSIMILATING AMERICA

Despite the growing realization that America was not Asia, some Europeans persisted in pursuing tropical paradises and beasts, Amazons, cannibals, and other perversions of nature. With its dense forests and great rivers, Brazil, some European writers reported, was inhabited by men with eight toes and feet pointing backward, people with dogs' heads and one eye, a tribe of half-men half-fish, and a society of warrior women who had no need of men. Fantastic Brazil was contradicted by realities on the ground in the encounter with natives, who like humans fought against the intrusion of foreigners who came "to occupy the land of Brazil, and were interested in its native men only as labourers and women as concubines."[21]

Other Eurocentrisms might have tilted representations of wondrous Brazil and its natives toward views of the noble savage, a cousin of childlike innocents in paradise. The Dutch, in their animosity toward Spain, which for a period before 1609 had controlled the Low Countries, derided Spain's ostensible "civilizing" mission of "savages" in America. Instead, some Dutch intellectuals maintained, America was not a "new" world but an ancient land with civilizations that comprised states organized around civic virtues, thriving markets, and intricate arts and crafts that long predated the arrival of the Spaniards. From 1570 to 1600, that "process of geographic assimilation," the notion of Dutch kinship with America's "Indians," both having suffered under the yoke of Spanish tyranny, one scholar noted, advanced the interests of the evolving Dutch republic and its strivings for nationhood and served "to domesticate, not exoticize," America and its peoples.[22]

Those nationalisms at home influenced the course of extravagant empires abroad such that Spain and, after its seizure by Philip II in 1580, Portugal and their lucrative trade routes and commercial outposts in Africa, America, Asia, and the Pacific became fair game for Dutch poaching. Portuguese Brazil, with its strategic location and valuable tropical products, especially sugar, offered a tempting target for the Dutch, who mounted expeditions against the colony in 1624 and 1630.[23] In 1636, the Dutch West India Company dispatched twelve ships and 2,700 men under the command of Johan Maurits, count of Nassau-Siegen, as governor of its Brazil colony. Among his expeditionary force were scholars, scientists, artists, and craftsmen whose assignment was to discipline unruly, tropical Brazil, which, upon landing, the count declared to be "a most beautiful world." During his seven-year rule, Johan Maurits built towns and roads, bridges, and spacious residences complete with aviaries and botanical and zoological gardens well stocked with tropical plants and animals.[24]

Complementing and advancing the governor's efforts to rule over and domesticate wild Brazil were his assembled scientists, like Willem Piso, who studied tropical diseases, and Georg Macgraf, who collected specimens, drew maps, and tracked celestial bodies and heavenly phenomena; and artists, like Albert Eckhout and Frans Post, whose paintings and sketches offered some of the first visual glimpses of landscaped "America," as they projected it, to eager European audiences and collectors.[25] Dutch paintings of a teeming, lush, and alluring Brazil gained popularity within the wider circuits of European desires for tropical exotica and products, as manifested in the famous zoological and botanical gardens of Leiden and Amsterdam and other European cities and estates such as the Versailles gardens of France's Louis XIV. Those collections of "curiosities," including the natives of Africa, America, Asia, and the Pacific, tropical plants, and live and stuffed fishes, birds, and mammals, elevated the social status of their owners as men of science, wealth, and the world.[26] Twenty-seven of Frans Post's Brazilian landscapes were a treasure trove, accordingly, when gifted to Louis XIV in 1679.[27]

Perhaps it was Post's brother, Pieter Post, a famous architect, who recommended him for Maurits's 1636 expedition to Brazil. Whereas scientist Georg Macgraf expressed his intention to "not write about anything which I have not actually seen and observed," Dutch portraits of America, including Post's renderings, depicted the New World through the eyes of the Old, the unfamiliar through the familiar. Post's Brazilian landscapes, thus, present compositions evocative of the European countryside but filled with a host of exotic plants, animals, and peoples, a veritable museum catalog of rare and exemplary specimens.[28]

In another act of assimilation and conversion, the crown and church recognized the humanity of Americans when Queen Isabella of Spain suspended American slavery and in 1500 declared the natives of America to be "free and not subject to servitude" and a 1537 papal bull acknowledged that "the Indians, as veritable men . . . , can in no way be made slaves or deprived of their goods."[29] Still, pious pronouncements of native humanity failed to contain the contagion of slavery or brutality in America, frequently justified on the grounds of racial inferiority among those lacking, in Jean Bodin's terms, "inner heat." America's tropics naturalized and rendered commonsensical the servitude of its inhabitant monsters, cannibals, Amazons, and concubines to European men.

Those medieval ideas of climate, bodies, and their merits, an ancestry rooted in ancient Greece, found new expression in the Enlightenment. About the time of Blumenbach's thesis on climate's influence over bodily conformations and "varieties" of humankind, eighteenth-century German philosopher Immanuel Kant claimed that in "hot countries the human being . . . does not . . . reach the perfection of those in the temperate zones. Humanity is at its perfection in the race of the whites." Furthermore, "the inhabitant of the temperate parts of the world, above all the central part, has a more beautiful body, works harder, is more jocular, more controlled in his passions, more intelligent than any other race of people in the world." In the same

FIGURE 1. *Brazilian Landscape with Anteater* (1649) by Frans Post. The tranquil and well-watered countryside is framed by a profusion of fecundity, tropes of the womanly tropics, including the pineapple, palm, and banana, all heavy with fruit. Naturalist Carl Friedrich Philipp von Martius considered Post's detailed renditions worthy of scientific illustrations. Source: Munich, Alte Pinakothek.

period, G. W. F. Hegel believed that "historical peoples" could not be found in the frigid or torrid zones, and that the "true theatre of history" was enacted only in the temperate zone, notably in its northern part, where "free movement" was possible.[30]

TROPICAL INVENTION

Like Columbus and Vespucci, both subjects of his studies in later life,[31] Alexander von Humboldt surveyed Spain's America, and between 1799 and 1804 his renderings of the physical world and its biotic communities "helped invent the tropics both as a field for systematic scientific enquiry and a realm of aesthetic appreciation," with art and literature being indispensable to science for a comprehensive understanding of nature, according to one historian.[32] The German naturalist, peering through European, Romantic spectacles, saw order, abundance, and fecundity in his New World, affirming his faith in an essential unity of knowledge and the cosmos, a creed fabricated in the Old World.

"The natural sciences are connected by the same ties which link together all the phenomena of nature," Humboldt wrote in the introduction to his monumental *Personal Narrative of Travels to the Equinoctial Regions of America* (1814–25). And tropical America was an ideal place to prove his beliefs. "America offers an ample field for the labours of the naturalist," he explained. "On no other part of the globe is he called upon more powerfully by nature to raise himself to general ideas on the cause of phenomena and their mutual connection. To say nothing of that luxuriance of vegetation, that eternal spring of organic life, those climates varying by stages . . . , and those majestic rivers." Amidst such grandeur and activity, the "savages of America" fade into the background, and, like descriptions of island paradises and their natives "in whose character we find a striking mixture of perversity and meekness," depictions of their station of "half-civilization" merely lends "a peculiar charm" to those of their deviant, feminized cultures.[33]

An "ardent desire," Humboldt confessed, fired his infatuation with the tropics and prompted his expeditions to "distant regions, seldom visited by Europeans," to places of "irresistible attraction" and "positive danger." In isolated and sedentary Europe, those "pleasures" of the "luxuriant regions of the torrid zone" held "a fascinating power." The contrast between worlds began with the skies, Humboldt related, when as he crossed the equator the familiar, reassuring constellations disappeared into the exotic, southern night and hemisphere. "Nothing awakens in the traveller a livelier remembrance of the immense distance by which he is separated from his country, than the aspect of an unknown firmament," he observed. And though losing one's bearings may be disorienting, the prospects of discovery and sightings of new formations excite appetites and passions.[34]

Those reveries and the mild climate and ocean calm could not eliminate the perils of the tropics, "the germs of a malignant fever" that boarded the vessel in the excessive heat. "Two sailors, several passengers, and, what is remarkable enough, two negroes from the coast of Guinea, and a mulatto child, were attacked with a disorder which appeared to be epidemic." The fever abated with fresh air and rest, except for one victim, who died, and the experience redirected the expedition away from Mexico to a landing on Venezuela's coast.[35] "It is well known," Humboldt reflected, "that Europeans, during the first months after their arrival under the scorching sky of the tropics, are exposed to the greatest dangers." The ability to acclimate, he proposed in the dominant discourse of his time, appears related to the "difference that exists between the mean temperature of the torrid zone, and that of the native country of the traveller, or colonist, who changes his climate; because the irritability of the organs, and their vital action, are powerfully modified by the influence of the atmospheric heat."[36]

Humboldt, from the ship's deck gazing toward the shore the morning of July 16, 1799, presented a spectacle that became the period's model for "the equinoctial regions." "Our eyes were fixed on the groups of cocoa-trees which border the river: their trunks, more than sixty feet high, towered over every object in the landscape," he detailed with the eyes of a scientist and artist.

> The plain was covered with the tufts of Cassia, Caper, and those arborescent mimosas, which, like the pine of Italy, spread their branches in the form of an umbrella. The pinnated leaves of the palms were conspicuous on the azure sky, the clearness of which was unsullied by any trace of vapour. The sun was ascending rapidly toward the zenith. A dazzling light was spread through the air, along the whitish hills strewed with cylindric cactuses, and over a sea ever calm, the shores of which were peopled with alcatras [a brown pelican], egrets, and flamingoes. The splendour of the day, the vivid colouring of the vegetable world, the forms of the plants, the varied plumage of the birds, everything was stamped with the grand character of nature in the equinoctial regions.[37]

Notably absent along the shores "peopled" by birds were America's human natives. When local inhabitants do appear, they are typically naked and lazy, acclimatized as they are to the simple monotony and enervating heat of the tropics. "The copper-coloured native," Humboldt reported, "more accustomed to the burning heat of the climate, than the European traveller, complains more, because he is stimulated by no

FIGURE 2. *Morning in the Tropics* (1877) by American artist Frederic E. Church, one of the foremost painters of the Hudson River school and shaper of the image of the tropics in the American mind. Natives, except for their material artifacts, fail to despoil this sublime vision of Paradise. Humboldt's accounts inspired Church to retrace his steps in tropical America. The Hudson River school mirrored the major themes of U.S. history—of discovery, exploration, and colonization—and aspired to scientific accuracy through direct observation and to revelation of God's designs in the natural world. Source: National Gallery, Washington, D.C.

interest. Money is without attraction for him." Further, he added, tropical abundance stifles industry and human creativity. "The richness of the soil, and the vigour of organic life, by multiplying the means of subsistence retard the progress of nations in the paths of civilization. Under so mild and uniform a climate, the only urgent want of man is that of food." Thus, in a narrative reminiscent of the ancient Greeks, "we may easily conceive why, in the midst of abundance, beneath the shade of the plantain and bread-fruit tree, the intellectual faculties unfold themselves less rapidly than under a rigourous sky, in the region of corn, where our race is engaged in a perpetual struggle with the elements."[38]

A more contemporary, Romantic sentiment enters Humboldt's text, nonetheless, in the form of the noble savage, the native whose tools barely scratch the soil's surface "as a transient guest, who quietly enjoys the gifts of nature."[39] Although such a "man" is to be admired, he fails to assemble in large cities, harness the environment, or build complex and great civilizations. In the tropics, the German scientist postulated, wild, "spontaneous" vegetation will predominate over "cultivated" plants. Likewise, its un-tutored, feminized "men" will require long periods of evolution to abandon the state of nature for the heights of civilization and shoulder the manly burdens of "absolute master" or "empire" builder. That ascent parallels the native's inadequacy to exert do-minion over unruly, womanly nature. And in the highly charged language of sex and conquest, Humboldt described the temperate zone as soil cleared, implanted, and stim-ulated to yield fruit, whereas "the torrid zone will preserve that majesty of vegetable forms, those marks of an unsubdued, virgin nature, which render it so attractive and so picturesque." "In proportion as we penetrated into the forest," the naturalist wrote revealingly, "the barometer indicated the progressive elevation of the land."[40]

While creating tropical illusions, Humboldt appears confounded by the equally chimerical depictions of Columbus and other European "explorers" of America as Asia. Europeans, he noted, saw America as "the advanced capes of the vast territories of In-dia and eastern Asia." Narratives by Marco Polo, Mandeville, and others advertised the immense wealth of the Orient in gold, diamonds, pearls, and spices, and Colum-bus, "whose imagination was excited by these narrations, caused a deposition to be made before a notary, on the 12th of June, 1494, in which sixty of his companions, pilots, sailors, and passengers, certified upon oath, that the southern coast of Cuba was a part of the continent of India." The treasures of Cathay and Japan, "which had fired the admiral's ambition in early life, pursued him like phantoms in his declining days." Those "geographical illusions" and that "mysterious veil" fixed Europe's atten-tion on the New World and made more insistent its pursuit of wealth in America. Unfortunately, those prospects went largely unrealized, Humboldt found, and many areas of promise had "relapsed into a savage state," affirming "the strange and some-times retrograde nature of civilization in America."[41] Humboldt's contemporary, Ger-man geographer Carl Ritter, called him "the scientific re-discoverer of America,"[42] in a gesture to Columbus, for having lifted the mythic, Asiatic veil.

Humboldt, like many European "travelers" before him, pursued figurative and

FIGURE 3. Frontispiece to Alexander von Humboldt, *Atlas géographique et physique du Nouveau Continent* (Paris, 1814), bearing the caption, "Humanitas, Literae, Fruges" (Civilization, Letters, Grains), gifts of the Old World to the New, showing Minerva (goddess of handicrafts) and Mercury (god of commerce) lifting a fallen Aztec prince and, in effect, the upended bust of an Aztec goddess. Humboldt's version was a modification of a drawing by François Gérard, a prominent painter of the French empire, whose depiction was captioned, "America, raised from its ruins by industry and commerce."

literal veins of gold in his version of "equinoctial" America. Production and commerce, he noted in his invention of the tropics, were lodged in plantations of sugar and slaves, although he considered African enslavement a vile institution. "A free, intelligent, and agricultural population," he predicted, "will progressively succeed a slave population, destitute of foresight and industry."[43] In addition to the green gold of sugar plantations, Humboldt's tropics hold the promise of mineral wealth, not unlike the original plunder of America's silver and gold by Spanish conquistadors. His lengthy descriptions of America's geologic formations, the former inspector of mines admitted, were "calculated to excite" interest in the tropics and prompt the pouring of European investments, "vast sums of money," into mining activities in Argentina, Chile, Colombia, and Mexico.

In sum, the tropical hermeneutics advanced by the German naturalist and a chief inventor of the tropics was a science of development and improvement, the endowments of the European temperate to the American tropical zones. "Metals are a merchandize purchased at the price of labour, and an advance of capital," Humboldt argued, "thus forming in the countries where they are produced, a portion of commercial wealth; while their extraction gives an impetus to industry in the most barren and mountainous districts."[44] Unaccounted for in that equation, like the rendered invisible indigenous peoples, were the gains accumulated by the imperial centers in the temperate zone and deficits by their hinterlands in the tropics.

A mature Charles Darwin attested to the power of Humboldt's tropics when he wrote, "My whole course of life is due to having read and re-read as a youth his 'Personal Narrative.'" And although disappointed that the German naturalist's work fell short of his scientific ambition to encompass the entire natural world, Darwin called him "the greatest scientific traveller who ever lived."[45]

ENCHANTED ISLES

Humboldt's moves were not random, as was revealed in his investments in metals, or mere tours of distant climes, as evoked by his aesthetics and sensibilities; they were active agents of Europe's empires. Voyages from the temperate homeland to the

FIGURE 4. Title page by Christopher Switzer for John Parkinson's *Paradisi in sole paradisus terrestris* (London: Lownes and Young, 1629). Although his sin drove him from the Garden, Adam served God by tending his garden for pleasure and necessity. Amidst gigantic flowers rises a magnificent pineapple.

tropical fringes and conquests of waters, lands, peoples, and plants were neither uni-lateral nor absent resistance, nonetheless. In unpredictable formation were nation-states, economies productive of surpluses, scientific and technological advances, em-pires, and the fickle rise and mix of individuals, groups, and resources. And yet, revealed in history's designs, insofar as themes emerge, are the workings and calculations of ac-tors who build upon, disassemble, and reconstitute toward certain ends. European botan-ical gardens in this era of empire and expansion help to ground those abstractions.

The earliest treatise on architecture may have been a first-century B.C. work by Marcus Vitruvius Pollio, discovered and first published in 1486. Vitruvian theory, pre-dominant over the next two centuries, stressed "the ideals of unity, proportion, and symmetry; the belief that geometrical forms revealed the divine order hidden in na-ture; and the decorum that assigned certain kinds of style to correlated kinds of build-ing."[46] That idea of divine revelation in Christian medieval Europe found later ex-pression in botanical gardens, the first planted in Oxford, England, in 1621, which strived to recreate the biblical Garden of Eden or a Paradise on earth. Perhaps the in-spiration for that conceptual use of space arose during the Middle Ages from an en-closed area adjacent to church buildings called paradise, which was designated for med-itation and prayer. Similarly, communicants were said to have entered Paradise, and the main body of the church was referred to as Paradise. Located geographically, Par-adise lay to the east of the torrid zone in the Southern Hemisphere—or so it was widely believed from claims made by the likes of Mandeville and the myth of Prester John, an Oriental Christian king of great beneficence and wealth.[47]

Ambition and power, an exhibit catalogue observed, motivated prints of elab-orate "European pleasure gardens" from the Renaissance to the age of revolution. Intended mainly as notices of wealth and ostentation, gardens and their reproduc-tions announced as well as documented their owners' estimation of their worth, civility, and grace. Vitruvian truths informed the designs of European gardens and houses of the elite—a style that began in Italy and reached its height at Versailles, to which Louis XIV, the "Sun King" and model of an absolute monarch, removed in the late seventeenth century. Away from the clutter of Paris, the gardens and build-ings of Versailles were integrated into a unifying whole, inducing "pleasure" in that "enchanted isle," a vision of Paradise where science and the senses converged and

FIGURE 5. *Veue du Chasteau de Versaille* (View of the Chateau of Versailles) by royal printmaker Israël Silvestre (1673). By 1690, Versailles, with its 230 acres of gardens, 1,730 acres of an inner park, and 16,000 acres of an outer park, surpassed all other palaces and gardens in Europe and was designed to epitomize the peerless magnificence of the French court and culture. Source: Elizabeth S. Eustis, *European Pleasure Gardens: Rare Books and Prints of Historic Landscape Design from the Elizabeth K. Reilley Collection* (New York: New York Botanical Garden, 2003), 34.

gained prominence even as nature, divinely ordered, and buildings, man's creations, lost their distinction.[48]

European gardens came to mirror the rise of systematic science in the naming and classifying of plant and animal life, a procedure that claimed man's mastery over nature.[49] Garden designs commonly featured the four continents and their plants imported from the colonies, another display of power during the age of European expansion. Gardens, thus, like museums and zoos, were the receptacles of empire, housing collections of exotic specimens from the tropics for study, preservation, and, in the

FIGURE 6. Cover of John Parkinson's *Theatrum Botanicum* (1640). The "Theater of Plantes" apprehends the entire world represented by the four continents: Asia on her rhinoceros, Europe in her carriage, Africa on her zebra, and America on her llama, along with their plants. Among America's distinctions is the pineapple.

instance of living forms, propagation. Hans Sloane, physician to Jamaica's British governor, gathered and drew more than eight hundred species of plants during his two years in the islands and introduced them to the British reading public in his *Voyage to the Islands of Madera, Barbados, Nieves, S. Christophers and Jamaica* (1707–25). Such registers of climate, flora and fauna, peoples, and diseases by naturalists helped to colonize new worlds, and the science of acclimatization, which strived to move and breed species outside their "natural" environments, consorted with the desire to exert authority over foreign lands and resources, enabling the political and economic transformations in and transfers among the peripheries and their centers.[50]

BOTANICAL EMPIRE

Notably, the eighteenth-century empire of plant science was heavily endowed with its applied value, or the utility and economy of plants, as is revealed in the life and labors of Joseph Banks.[51] Like the merchant capitalists of their day, "plant mercantilists" searched the tropics for products with value for trade to enrich themselves and their companies, and their network of collectors in the field, gardens in the colonies, and collections and laboratories at home were as much imperial as the implantation of European flags and offices around the globe. Eighteenth-century usage of the word *botanist,* in fact, designated an "enquirer into the nature and property of vegetables [who] ought to direct his view principally towards the investigation of useful qualities."[52] Banks was one of the age's most successful botanical entrepreneurs.

Heir to a large fortune and Oxford educated, Banks gained fame by joining James Cook's first expedition (1768–71), which sailed to the South Pacific and made stops in Tahiti and New Zealand. Banks took with him two artists, two naturalists, four servants, two of his pet dogs, and an enormous amount of baggage. For his heroics, King George III awarded Banks the Order of the Bath, while cartoonists mirrored the widespread doubt about the importance of Banks's collection and drawings of plants and animals.[53] Upon his return to England, Banks enjoyed celebrity status, dining with Samuel Johnson and others of London's educated elite and dominating the talk of the town, eclipsing even Cook.

When the Admiralty prepared a second voyage to the South Pacific in 1771, the press referred to it as "Mr. Banks' voyage" despite its being led by Cook, and Banks was put in charge of the natural history portion of the expedition. This time Banks planned an even larger party of collectors, painters and draftsmen, secretaries, six servants, and two horn players for entertainment. He also convinced Cook to alter the ship's interior to make the cabins more spacious and persuaded Parliament to allocate funds for James Lind, physician and astronomer, to join his team. Before departure in May 1772, Banks held a festive party on the *Resolution*'s deck for a few guests, including John Montagu, the Earl of Sandwich and First Lord of the Admiralty, and the French ambassador, complete with servants and a small orchestra. After a trial run, Cook decided the Banks-inspired alterations rendered the ship ill suited for the long voyage and ordered that the *Resolution* be returned to its original state. Incensed, Banks quit the expedition and never again sailed the Pacific.[54]

Banks was not, however, finished. Even though he was not a member of Cook's second voyage (1772–75), the expedition collected specimens for him, and he remained the key figure in the accumulation and distribution of natural history materials from all three of Cook's voyages.[55] On the first Cook expedition, he had selected the seeds and plants from Europe and America to take to Pacific lands to cultivate, thereby effecting a significant distribution of plants and directing the economic, both subsistence and commercial, life of the colonies. He continued that work of accumulating, testing, and redistributing from his London study and the Royal Botanic Gardens at Kew, which Banks helped build and where he and others conducted experiments to ascertain the climatic limits of plants.[56] And Kew's satellites—botanical gardens on Jamaica and St. Vincent in the West Indies, St. Helena in the South Atlantic, and in Calcutta and Madras in India, and Sydney and Parramatta in Australia—all bore Banks's imprint.[57] Moreover, Banks possessed a vision of empire based upon the science and economics of plant transfers. "Scientific knowledge coupled with enterprise and industry could be utilised to augment the biological resources of the British colonies for the aggrandisement of the mother country," Banks held. But he also claimed that of first import as beneficiaries of those transfers were the tropical colonies.[58]

From 1770 to 1820, some 126 collectors outside of Europe served Banks or Kew Gardens, and, according to one study, "there was scarcely a part of the world unrepresented in the travels of the Banksian collectors. . . . Banks's voracious appetite for botanical specimens was ecumenical."[59] Banks, albeit extraordinary, was simply one of many agents of Europe's botanical empire, which effected plant and animal transfers to sustain colonies and economic systems such as the sugar plantations in the West Indies. When the islands appeared incapable of feeding their enslaved workers, the British took from the South Pacific breadfruit trees to plant in the Caribbean to alleviate that perceived shortage.[60] As a historian of science observed, "The botanical sciences served the colonial enterprise and were, in turn, structured by it. Global networks of botanical gardens, the laboratories of colonial botany, followed the contours of empire, and gardens often served its needs." By the end of the eighteenth century, she noted, Europeans had planted 1,600 gardens around the world, and scientists like Banks "reigned over metropolitan botanical gardens . . . , sitting like anointed monarchs at the centers of vast botanical empires."[61]

Along with cash crops such as cotton,[62] Europe's botanical empire was centrally concerned with the culture of herbs and plants for their medicinal properties. The ancient Greeks developed a systematic knowledge of medicinal plants, as in Dioscorides' *De materia medica,* which was still being used in the seventeenth century, and in the sixteenth century "physic" gardens were attached to hospitals and universities in Italy. Those spread to other parts of Europe, and medical training involved instruction in natural history. Kew Gardens and Banks, accordingly, sent out collectors who were also physicians, and while they tended to the ailments of imperial troops and merchants in tropical China, India, and America they searched for botanical cures. In those wildernesses, confronted with the spectacle of abundance but also chaos, Banksian agents strived to transcend confusion and "reduce the natural world of empire to order."[63] After all, science presumed mastery over nature and society, conquering climates and diseases and solving intractable social and economic problems. And that work could be done only by Europeans, they assumed, mired as other peoples were in darkness and stagnation.

Stripped of its beasts, Amazons, and other perversions of nature, the tropics gifted Europe with precious metals and green gold from its plantations and gardens. In its transfer of plants across the tropical and temperate zones, European science violated "nature," previously believed to have been divinely ordained, and conspired to institute and maintain a global, European empire. That rational economic and botanical

HISTOIRE
NATVRELLE ET MORALE
Des
Iles Antilles de
l'Amerique

FIGURE 7. Frontispiece to Charles de Rochefort's *Histoire naturelle et morale des Iles Antilles de l'Amérique* (Rotterdam, 1665). America's native peoples and exotic animals and plants surround a seated European monarch. Pineapples appear, cradled in the hands of a kneeling supplicant and in the garland above.

order, historian David Mackay proposed, reaped "immense benefits to Britain and her empire" but also functioned to affirm that "European reason might even produce order in the realm of humankind."[64] In that additional sense, in its claim to install order amidst chaos, European botanical science, like its allies in history and geography, with its observations, classifications, and quantifications often in disregard of indigenous knowledges, properly belongs to the taxonomy of empire.

3

Tropical Fruit

"It is just about fifty-five years since a boy of fifteen stood on the docks at Boston and gazed with wondering eyes at the ships unloading their freight from all parts of the world," a 1916 article in a popular American magazine began. "But as he stood day after day watching the dock scenes he was most attracted by the ships that came from the tropics, loaded with fruits, coffee, and spices. There were few of them, for in those days the tropical countries were perilously distant and unhealthy; but from the sailors who had braved the Caribbean the boy heard stories that filled him with wonder." That carefully crafted tale, "these romantic details in the life of one of the influential industrial leaders of the time," recounted how Andrew W. Preston, Massachusetts born in 1846, became "an industrial empire builder" as president of the United Fruit Company. The portrait profiled a great humanitarian and visionary: "And then this fifteen-year-old boy dreamed of going to the tropics and wiping out the scourge of disease and organizing communities that would send to all ports of the world great ship-loads of fruits, which the sailors said the tropical lands were capable of producing."[1]

Another, less laudatory rendering of the United Fruit Company, which in the worshipful language of a supporter brought "the tropics and the United States into closer industrial communion," depicted a firm and a "banana empire" that "throttled competitors, dominated governments, manacled railroads, ruined planters, choked cooperatives, domineered over workers, fought organized labor, and exploited consumers. Such usage of power by a corporation of a strongly industrialized nation in relatively weak foreign countries constitutes a variety of economic imperialism." One of this essay's authors, a United Fruit Company employee for fifteen years in Costa Rica and Panama, reported that the "banana empire" was "not primarily an aggregation of mutually interacting governmental and industrial agencies, but the expansion of an economic unit to such size and power that in itself it assume[d] many of the prerogatives and functions usually assumed by political states." Rather than rely upon the U.S. State Department, the study revealed, the United Fruit Company trained its own representatives to deal directly with governments in the Caribbean and Central America.[2]

The United States was "conceived in imperialism and dedicated to the principle of expansion," declared Harry Elmer Barnes in his introduction to a book on the "banana empire."[3] The expanding frontier from colonial Spanish Mexico northward and, later, the Atlantic seaboard westward involved the conquest of indigenous peoples and the continent. When founded, the new nation fulfilled its self-proclaimed "manifest destiny" in westward expansion, and by 1890, with its interior occupied, it embarked upon an overseas mission in search of new frontiers. Moreover, Barnes observed, that "new national imperialism" constituted a mere fraction of a vast and rapid global spread of European peoples. "In 1800 about four-fifths of the land area of the world had not been opened to civilized man through exploration," he wrote, "and as late as 1870 more than half of the habitable surface of the earth had not been penetrated by Europeans. By the beginning of the twentieth century the whole planet outside of the extreme polar regions had been traversed by the white man and its resources and potentialities for exploitation had been catalogued."[4] Those movements and mappings plotted and archived European desires.

TROPICAL DISEASES

The ventures of explorers and traders to the edges of their known world were calculated for the returns won in the course of those outward travels. Some of those gains included representations of places and peoples and the acquisition of goods and objects and animal and human specimens. Passengers, seen and unseen, lurked among those newly and sometimes ill-gotten possessions. Migrant workers from the tropics were indispensable for the factories and fields of the temperate zone, but those who worried over the rising tide of "alien" and "inferior" stocks compared their inward movements to bacterial invasions. Invisible, silent, and deadly, pathogens and diseases

purchased in contact and intercourse with the natives of the tropical band, like undesirable "races" and racialized contagions, required defenses such as exclusion laws
and quarantine stations. That realization dawned early on the foreigners from Europe
who, with their guns and germs, encroached upon and, in part, visited pestilence upon
the New World, both continent and islands.[5]

European medicine of the seventeenth and eighteenth centuries generally adhered
to the Hippocratic thesis that climate and environment played a significant if not absolute role in determining constitutions and health. So when ships from the West Indies, where an epidemic was rumored, arrived in Boston in 1647–48, the Massachusetts General Court ordered a day of fasting, "having cause to feare least our sinnes
may provoke the Lord to lay more heavy corrections upon us," and ordered a quarantine on those ships. In the mid-eighteenth century, the Carolina General Assembly, aware that smallpox, yellow fever, and other diseases were endemic to Africa, required all slaves imported from that continent to spend a ten-day quarantine on
Sullivan's Island at an edge of Charlestown harbor.[6] Other port cities such as New
York, Philadelphia, and New Orleans followed that practice of quarantine. Widely
opposed by merchants whose ships and cargo were held hostage in home anchorages,
quarantines were deployed because the tropics were, in the words of the Andrew W.
Preston and by extension the United States epic, "perilously distant and unhealthy."

In July 1793, amidst Philadelphia's summer heat, "fleets of ships came in from the
West Indies, discharging from their crowded holds great hordes of refugees, white,
black, mixed, from the French island of Santo Domingo," according to a history of
Philadelphia's "great plague." "Gaunt, hungry, sickly, they poured into the city, bringing news of a great revolution in the sugar islands, of a horrible carnage and slaugh-
ter, of the destruction of towns and the ruin of merchant houses."[7] Colonized by Spain
and then France, Santo Domingo (or Hispaniola) was built and made prosperous
through the labor of enslaved Africans whose 1791 rebellion and its aftermath
prompted the refugee stream to Philadelphia. The landing was a return of sorts in that
Pennsylvania troops had fought in the West Indies in the 1740s and some of Philadelphia's wealthy merchants had made their fortunes in trade with Santo Domingo.[8] The
tropics, after all, were not all that distant, though "unhealthy."

The refugees numbered more than two thousand by August, when the city totaled about fifty thousand, and their arrival sparked an outbreak of yellow fever among white and black residents alike. Hundreds were taken ill, dozens died in September, and panic seized the city, despite the fact that virtually every port city along the East Coast and New Orleans had experienced yellow fever attacks since the 1690s. A Philadelphia physician, William Currie, described the fever's course: a chill, headache, and "soreness at the stomach" for the first few days, "immediately succeeded by a yellow tinge in the opaque cornea . . . black vomit . . . haemorrhages from the nose, . . . agitation, deep and distressed sighing, comatose delirium and finally death."[9] Physicians debated the source of the contagion. Some, like Benjamin Rush, believed it to derive from the city's putrid air, while others, like John Lining, favored importation as the explanation. Some white leaders accused blacks of profiteering from the plague, leading to a response from prominent African American Philadelphians Richard Allen and Absalom Jones, founders of the Free African Society in 1787.[10] In the end, probably more than five thousand died of the disease, and the 1793 plague became one of the most written-about episodes in U.S. medical history.[11] In addition, the yellow fever epidemic, which resulted in thousands of deaths in East Coast cities during the 1790s, led to sanitary campaigns and a public health movement initiated by volunteers and government to create health boards, hospitals, and welfare programs.[12]

Yellow fever became better understood in the field in the act of empire creation. In 1898, the United States invaded Spain's possessions in the Caribbean and Pacific, and in the course of that war and the occupation that followed, hundreds of soldiers died of yellow fever, especially in Cuba. As a result, in 1900 the U.S. secretary of war formed the Yellow Fever Commission to Cuba, which was the fourth commission to study the malady, and directed it to pursue "scientific investigations with reference to the infectious diseases prevalent on the Island of Cuba."[13] The commission, headed by Walter Reed, tested the hypothesis of Cuban-born Carlos Finlay that mosquitoes transmitted yellow fever by injecting the blood of an infected person into the body of an uninfected person. At first, commission members experimented on themselves, and later they employed volunteers who signed informed-consent documents—the first instance of such documents' use in human research. The results

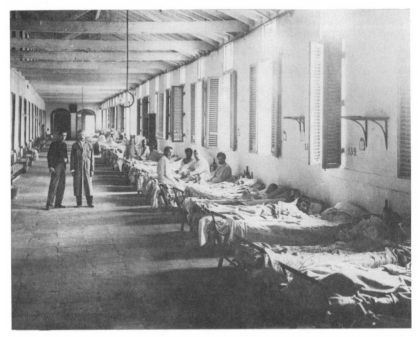

FIGURE 8. Patients in a yellow fever hospital, Havana, Cuba, c. 1899. Courtesy of Library of Congress.

proved Finlay's theory, and the commission forthwith recommended a mosquito eradication campaign to eliminate the threat of yellow fever in Cuba.[14] More famous, a similar mosquito eradication campaign enabled work on the Panama Canal, begun in 1901, and that "conquest" of a tropical disease was responsible for, from an imperialist perspective, "the advance of commerce and civilization, [and] the rejuvenation of Latin America."[15]

A triumph of "tropical medicine," the Cuba Commission's "rigorously conducted and controlled experiments . . . changed hygienic policy almost overnight and eliminated yellow fever as a serious public health threat from Cuba in a matter of months," declared a president of the American Society of Tropical Medicine and Hygiene, a body formed in 1903.[16] Meanwhile, the United States annexed Puerto Rico in 1898, occupied Cuba from 1898 to 1902 and in subsequent years, exercised "titular sovereignty" over Panama's Canal Zone in 1903, assumed the customs collection of the Dominican Republic in 1904, sent 2,600 troops to Honduras and Nicaragua in 1906, and intervened in Honduras in 1907. In the Pacific, the United States supported the overthrow of the Hawaiian kingdom in 1893, annexed Hawai'i, the Philippines, and Guam five years later, and acquired American Samoa in 1899. As a chronicler of the American Society of Tropical Medicine and Hygiene observed without a trace of deception, "It is no coincidence that our society was founded at the very moment when the USA first emerged as a global power."[17]

During this period of empire, Philadelphia's Jefferson Medical College and the Polyclinic and College for Graduates in Medicine offered for the first time courses on "tropical diseases," a term first coined in 1787, and the Harvard Medical School established the first department of tropical medicine in the country.[18] The completion of the Panama Canal would lead to more intimate relations between the United States and the tropics, where "grave pestilential diseases" thrived, an announcement of Harvard's department predicted, and "as trade with these regions grows, the danger that these diseases may find entry into America will grow too."[19] A medical report elaborated: "Our comparatively recent acquisition of territory in the American and Eastern tropics gives us a renewed interest in the diseases that affect especially tropical regions." Not limited to the tropics, those diseases "occur sometimes under rather alarming circumstances in the temperate zones," and with an increase in commerce between those bands the contagion could no longer be contained. Sleeping sickness, once believed to be limited to "negroes" in Africa, could now be found in the United States, and the "oriental condition" of "running amok," where the afflicted "runs wildly through the street brandishing firearm or saber and killing every one whom he can reach," had spread to other places. Schools of tropical medicine, thus, helped to install effective colonial management and hygiene and sanitation at home, not to mention sanity.[20]

Geographers plotted the progress of tropical medicine. Summoning the account of Ronald Ross, a key figure in the establishment of British tropical medicine throughout the empire, a report published in the *Bulletin of the American Geographical Society of New York* praised "the first pioneers"—European explorers, soldiers, and colonizers— for their suffering in the far, "almost untouched" reaches of the tropical band. "With only poor huts to live in, nothing but the poorest food to eat and surrounded by innumerable insect-carriers of infection, they lived a most dangerous life. The risks of death from tropical diseases were immense. Whole regiments used to be wiped out by yellow fever in the West Indies and by cholera in India. No young man visiting tropical America or Africa could be reasonably certain of life." In addition, disease retarded the advance of natives, restricting the survivors to "a barbaric state." "But tropical medicine and bacteriology are doing away with these horrors," the article claimed, and the scourge of yellow fever, cholera, malaria, and dysentery has been tamed, saving both colonizers and the colonized and ensuring "civilization and prosperity for vast possessions in the tropics."[21]

Geography and medicine pursued the path of empire, from the insertions of "possessions" onto national maps to abetting the conquests of lands, peoples, and diseases.[22] Geography's idea of the climate's influence over its life forms, including human "races" and their temperaments and social organizations, as reflected in the "zonal locations" of Ellen Churchill Semple and "climatic energy" of Ellsworth Huntington, offered a baseline for tropical medicine. Can "races," acclimatized to their "locations," live and thrive outside of their "natural" habitats? was a research question posed by physicians tending the imperial troops in the torrid zone.[23]

Medical doctor and U.S. Army surgeon Charles E. Woodruff presented his opinion on that matter to the Manila Medical Society in 1904 as American troops were succumbing to Filipino bullets and tropical maladies. Naturalists have shown that species arise in "zoological zones" delimited by isothermals, wrote Woodruff, affirming the scientific maxim of his time. Humans, too, submit to that natural law, which is why we find "a separate type [race] in each zoological zone, and each type is unfitted for residence in any other zone markedly different from the ancestral one." So when "Aryans" leave their temperate homelands for the tropical sun, they expose their bodies to "light stimulation," which in turn can cause exhaustion ("tropical exhaustion"), stomach disorder ("tropical apepsia"), insanity ("tropical insanity"), neurosis ("tropical neurosis"), amnesia ("tropical amnesia"), and suicide ("tropical suicide"). Despite those debilitating effects of "tropical light on white men," the army surgeon concluded, worthwhile is the great cause of the age for Europeans and Americans, the conquest of the tropics "to give to its peoples that security of life and property, and that civilization and prosperity, which they cannot attain by their own un-aided efforts."[24]

WHITE MAN'S BURDEN

Reassuring and reinforcing were the allied scripts of empire builder Andrew Preston's life, the conventional narrative of U.S. expansion, and the beneficent mission of the entire project of European imperialism directed mainly at the tropical zone for its products and labor. As put by Mary Endicott Chamberlain, women of the Colonial Nursing Association served Britain, "to whose colonies it ministers" in restoring back to health "delicately nurtured women" who, in the colonies, were exposed to "suffering and peril" that were "inconceivable," little babies who "spend their brief lives in pain," and "strong men" struck down "by that baleful and deadly foe, tropical fever."[25] The U.S. president, William McKinley, explained to a group of Methodist ministers and missionaries his reasons for favoring the annexation of the Philippines and, in that act of assimilation, waging war against Filipinos who resisted first Spanish and then U.S. colonization.

> The truth is I didn't want the Philippines and when they came to us as a gift from the gods, I did not know what to do about them. . . . And one night it came to me this way—(1) that we could not give them back to Spain—that would be cowardly and dishonorable; (2) that we could not turn them over to France or Germany—our commercial rivals in the Orient—that would be bad business and discreditable; (3) that we could not leave them to themselves—they were unfit for self-government—and they would soon have anarchy and misrule over there worse than Spain's was; and (4) that there was nothing left for us to do but to take them all, and to educate the Filipinos, and uplift and civilize and Christianize them, and by god's grace do the very best we could by them, as our fellowmen for whom Christ also died. And then I went to bed, and went to sleep, and slept soundly, and the next morning I sent for the chief engineer of the War Department (our map-maker), and I told him to put the Philippines on the map of the United States, and there they are, and there they will stay while I am President![26]

FIGURE 9. Like the "before" and "after" photographs of Native American schoolchildren, these portraits of a Bontoc Igorot boy in 1904 and 1913 were intended to display the benevolence of American colonialism in the Philippines. From Dean C. Worcester, *The Philippines Past and Present,* vol. 2 (New York: Macmillan, 1914), 448.

The president's characterization of Filipinos, denizens of the tropics, as incapable of self-rule approached an intellectual tradition that extended at least as far back as Aristotle, who saw Asians as "by nature slaves" and Greeks "able to rule the world."

Not all Americans supported the president's ostensible uplift of Filipinos and other island peoples. Although some manufacturing industries may have been lured by the prospect of untapped markets in the tropical band, other business interests, such as sugar, saw unfair competition in the vast plantations and cheap labor of the islands. And though some Americans believed that the colored races were inescapably mired in barbarism, others held that they could receive instruction. But imperialists and anti-imperialists alike generally agreed upon the superiority of whites and inferiority of peoples of color.[27] Even within the gendered discourse of empire that underscored the manly nature of expansionism and conquest,[28] many suffragists took up the "white man's burden," exercising race privilege even as they were disenfranchised as domestic subjects. British Rudyard Kipling could, in 1898, exhort his fellow "Anglo-Saxons," white Americans,[29] to

> Take up the White Man's burden—
>> Have done with childish days—
> The lightly proffered laurel,
>> The easy, ungrudged praise.
> Come now, to search your manhood
>> Through all the thankless years,
> Cold, edged with dear-bought wisdom,
>> The judgment of your peers![30]

Reformer Anna Garlin Spencer, addressing the 1899 suffrage convention on the subject of "Our Duty toward the Women of Our New Possessions," paraphrased Kipling's popular verse to "feminize" the colonization of the Philippines:

> Take up the White Man's burden!
> Go, pilot as you may
> Your new-caught, sullen peoples
> Who do not know the way.
> Go, teach them not with cannon,
> Those fluttered folk and wild;
> But lead them, as, with yearning,
> A mother leads her child.[31]

Spencer lectured without recognition of the irony of her unreflective embrace of "our" new "possessions" when women in the United States were commonly considered the property of their fathers and husbands.

Such prominent suffragists as Susan B. Anthony opposed their country's war against Spain and its colonial subjects. Senator George F. Hoar of Massachusetts favored the war in Cuba but later characterized the protracted campaign in the Philippines as introducing "perfidy into the practice of war" and baffling "the aspirations of a people for liberty." In that war to "civilize" "our little brown brothers," more than

FIGURE 10. Having shouldered the "white man's burden," Uncle Sam assumes the education of his unruly children, the Philippines, Hawai'i, Puerto Rico, and Cuba, while an "educated" African American and Native American remain in their place and a Chinese appears at the door. "School Begins," *Puck*, January 25, 1899. Artist: Victor. Courtesy of Bishop Museum, Honolulu.

four thousand American troops and an estimated two hundred thousand Filipino soldiers and civilians died.[32] As writer Mark Twain imagined it, the befuddled Filipino might say: "There is something curious about this—curious and unaccountable. There must be two Americas: one that sets the captive free, and one that takes a once-captive's new freedom away from him, and picks a quarrel with him with nothing to found it on; then kills him to get his land."[33]

About two months before the war's official end in 1902, the New England Woman's Suffrage Association invited Clemencia Lopez to address the group. Lopez, member of the Filipino elite, was in the United States to plead for the release of her three brothers being held by the Americans in the Philippines. In her speech, Lopez equated the fight for Filipino nationhood with American women's fight for attainment of the vote. "We are both striving for much the same object," she reminded her audience, "you for the right to take part in national life; we for the right to have a national life to take part in." In fact, she concluded, patriotism in the Philippines "means that we must oppose the policy of yours."[34]

The American Anti-Imperialist League, formed in 1899, counted among its incongruous mix of members industrialist Andrew Carnegie, American Federation of Labor president Samuel Gompers, philosophers William James and John Dewey, social reformer Jane Addams, writer Mark Twain, and segregationist Senator Benjamin R. Tillman. Some believed in the league's platform that "all men, of whatever race or color, are entitled to life, liberty, and the pursuit of happiness"; others opposed the annexation of foreign lands as bad for business, labor, and race relations by the addition of inferior stocks into the nation. Like the tropical diseases that infiltrate and pollute, they warned, the "benevolent assimilation" of barbarous races would prove to be "perilously . . . unhealthy." Based upon the southern experience, "Pitchfork" Ben Tillman of South Carolina knew that the injection of "any more colored men into the body politic" and the mingling of "two races side by side that can not mix or mingle" would inevitably result in "deterioration and injury to both and the ultimate destruction of the civilization of the higher."[35]

EMPIRE'S WASTES

Global ambitions meant intercourse with the world's peoples, and the reciprocal of natives and markets abroad was alien bodies and diseases and exotic imports at home. In surveying the late nineteenth-century phase of U.S. expansion, Matthew Frye Jacobson astutely observed, "the foreigner abroad constituted only half of the sociopolitical equation under this regime of industrial progress and aggressive export; the resettled foreigner at home was the other."[36] And although the debate around imperialism and its brood—colonial subjects and immigrants—was an intensely national concern,[37] its reach and contours were emphatically transnational and global. Broader patterns emerge from those elevations of space and time; within those discourses, from the ancient Greeks to the late nineteenth-century European empires, are rank categories and valuations: temperate and tropical zones, white and colored races, civilized and savage states, and civilizations West and East.

The Rising Tide of Color against White World-Supremacy (1920) is one such text in that venerable tradition, published by Harvard Ph.D. and author Lothrop Stoddard. He and Madison Grant, trustee of the American Museum of Natural History and author of the best-selling *Passing of the Great Race* (1916), were among the leaders of the Eugenics Research Association, established in 1913 at Cold Spring Harbor, New York. Grant's introduction to *The Rising Tide* called Stoddard "a prophet" for boldly resurrecting the specter of the "Nordic" race's retreat and forecasting the potential for "a gigantic race-war." Simply turn to "the map, or, better, to the globe," he urged, and read history's course in "the light of geography."[38] The map, laid out by Stoddard as "a world of color," revealed whites concentrated in the temperate band north and south of the tropics, and yellow, brown, and black masses astride the equator and north to about the fortieth parallel.

Likening the swamping of the "inner dikes" of the white homelands by "a colored peril" to a contagion, Stoddard prescribed laws to isolate and quarantine the source of the invasion. Immigration historian Prescott F. Hall was correct, Stoddard agreed, when he wrote: "Immigration restriction is a species of segregation on a large scale, by which inferior stocks can be prevented from both diluting and supplanting good stocks. Just as we isolate bacterial invasions, and starve out the bacteria by limiting the area and amount of their food supply, so we can compel an inferior race to remain in its native habitat, where its own multiplication in a limited area will, as with all organisms, eventually limit its numbers and therefore its influence."[39]

The pattern of global migrations, Stoddard continued, that trespass upon and pollute white lands, was replicated in the United States. "The colonial stock was perhaps the finest that nature had evolved since the classic Greeks," he began, representing "the very pick of the Nordics of the British Isles." Here, in isolation on this continent, evolved, he quoted Madison Grant, "a pure race of one of the most gifted and vigorous stocks on earth, a stock free from diseases, physical and moral." But in the mid-nineteenth century, waves of migrations from northern Europe and then "the truly alien hordes" from southern and eastern Europe broke upon our shores and soon became a "veritable deluge." The Nordics augmented the colonial stock and ameliorated the danger, but the flood of darker mixtures still posed "a menace to the very existence of our race, ideals, and institutions."[40]

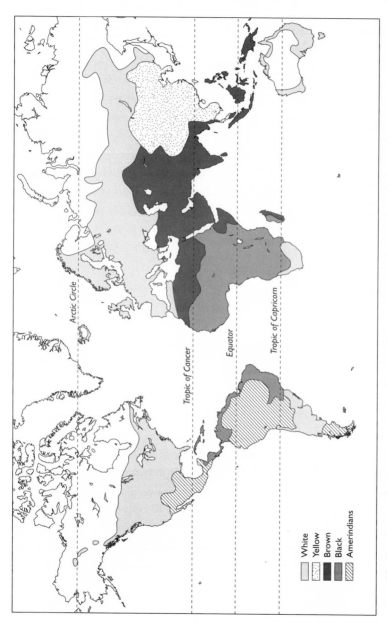

MAP 4. The "primary races" and their distribution as mapped by Lothrop Stoddard, showing the lines of battle for world dominion. Adapted from Lothrop Stoddard, *The Rising Tide of Color against White World-Supremacy* (New York: Charles Scribner's Sons, 1920).

White
Yellow
Brown
Black
Amerindians

Arctic Circle

Tropic of Cancer

Equator

Tropic of Capricorn

YELLOW PERIL

Although focused primarily upon "undesirable" European stock, in some of which coursed "Oriental" blood, the argument of those favoring immigration restrictions during the late nineteenth and early twentieth centuries prominently featured Asians.[41] In fact, expectation of an impending monumental struggle—the "yellow peril," probably named by Germany's Kaiser Wilhelm II in 1895—was widespread. The coming clash would be between East and West civilizations, religions, and military might.

The contest, perhaps as old as the wars between the ancient Greeks and Persians, took various forms depending upon the time and place. During the period of European and American empire, China and Japan, along with their mode of government and religion, were the principal obstructions to white supremacy. Even as East and West constituted polar extremes impossible of meeting, Oriental despotism and paganism were antithetical oppositions to Western civilization and Christianity. In the United States, China's immense and fecund numbers and diseases threatened to inundate and contaminate the body politic. Especially after its defeat of China in 1895 and Russia in 1905, Japan's modernized military and industries troubled the waters of the American Pacific lake.

A fitting symbol during this era of white supremacy and its contrived counter, the "yellow peril," was leprosy, which was thin epidemiologically but thick metaphorically. Prevalent in Europe and recognized by Christians as an abomination and a pollution and stigma, leprosy became, in the age of empire, a tropical disease of inferior peoples endangering conquests abroad and purities at home. "Our territorial expansion has created conditions favorable to the increase of leprosy in this country," an American physician told his fellow dermatologists in 1900. "The danger will come from the exposure of soldiers and sailors who form the army of occupation, of civilians who fill administrative posts in the government of these colonies, and of others who are attracted by considerations of trade and commerce and who are thus brought into intimate contact with their leprous population." Allegedly conveyed to and admitted into the United States by "slaves"—Africans and Chinese prostitutes and "coolies"— leprous and other diseased bodies demanded exclusion, isolation, and quarantine; their

FIGURE 11. Inscribed on this painting (1895) commissioned by Wilhelm II depicting the "yellow peril" were the words "Nations of Europe, defend your holiest possession." *Review of Reviews* (London), December 1895, 474–75.

presence "breeds moral and physical pestilence," said the president of the American Medical Association of Chinese prostitutes in 1876. The example of Hawai'i and its policy begun in 1865 of removing and segregating lepers to a remote spit of land on the island of Moloka'i became a model for the United States and the world. At that time in the Hawaiian kingdom, cases of leprosy averaged less than 150 each year among a total population of about 63,000, without evidence of its spread or increase.[42]

Actual infestations, like the 1876 San Francisco smallpox epidemic, which struck more than 1,600 and claimed nearly 450 lives, and the 1899 bubonic plague outbreak in Honolulu, inspired allegations of filth, crowding, and depravity, resulting in quarantines, burnings, and surveillance of Chinese and Japanese districts. Blame rested with the "unscrupulous, lying and treacherous Chinamen," charged physician John Meares, San Francisco's city health officer in 1876, and their "willful and diabolical disregard of our sanitary laws." In 1900, Hawai'i's Board of Health ordered a controlled burn of Chinatown, which spread out of control, destroying almost one-fifth of the city's buildings and leaving at least five thousand people homeless. The incidence of plague in Honolulu also prompted surveys along the West Coast of Chinatowns and Japantowns, the presumed "cesspools" and "spots" of racialized diseases and bodies, which had resisted confinement to their ancestral "native habitats" and transplanted themselves in temperate zones.[43] Quarantines such as the Moloka'i leper colony, "ethnic" enclaves in the United States, or the tropical girdle could not possibly contain a phantasm. "By the end of the nineteenth century," a study explained, "the imagery of leprosy [impurity] had merged with the symbolism of racism and yellow peril to the point where a distinction between the medical entity and unfavorable populations was effectively blurred."[44]

The "yellow peril" portended a host of threats, including the molestation of white girls and women, miscegenation and a "lowering" of white racial strains, homosexuality, bodily and spatial contagion and pollution, cheap labor and the displacement of white workers, unfair business practices, cultural and linguistic contamination, and political and military takeovers. Those dangers were detailed in popular books, magazines, and films, research surveys and scholarly studies, legislative hearings, classified intelligence reports, and security plans and military preparations. They were at once fanciful and real, and they achieved actualization in the perceptions and practices of citizens and aliens and in city ordinances, state laws, court rulings, and national immigration acts.[45] Enforcement of those imperatives included border patrols and holding centers, segregated neighborhoods, schools, and work sites, forbidden sexual couplings and marriages, and physical violence and expulsions. Asians and their landings in the United States and other white enclaves were a national and international "problem" from the perspective of late nineteenth-century white supremacy.

Written about the time of San Francisco's smallpox epidemic, which threatened expansion beyond Chinatown's confines, Last Days of the Republic (1879) by Pierton W. Dooner detailed the Chinese spread and ultimate conquest of the United States. The book, a "deductive history," purported to write American history for the twentieth century when a divided, complacent republic would fall to the Chinese "swarming horde." In line with an ancient Greek characterization of tropical, Ori-

FIGURE 12. Scene after the Honolulu Chinatown fire, January 20, 1900. The Kaumakapili Church, on Beretania at Smith Street, is surrounded by buildings reduced to smoldering rubble. Courtesy of Hawai'i State Archives.

ental peoples, Dooner's Chinese were "servile to the last degree, they seemed to be a people ordained by nature to be the servants of all mankind." But that surface belied an inner cunning, an "unwholesome spirit, seconded by a consuming avarice" and "a bold ambitious desire." China, with its "immense population of four hundred and seventy-five millions of souls," would under European and American prodding and tutelage awaken from its slumber and develop its industries, science, and military might, as was predicted of tropical peoples generally by Charles H. Pearson, whose influential book appeared some fourteen years later.

With its migrants an advance army monopolizing the labor force of a United States rent by political factions, China would seize control of the nation's industries, win political office through the ballot, and finally muster soldiers and whip the "native ferocity of the [Chinese] soldier into a furious and ungovernable flame of action." "Forever occupied and diverted by its factions and its politicians, in their local intrigues for the acquisition of political power," warned Dooner, "the Ship of State sailed proudly on, too blinded by her preoccupation and too reliant in her strength to bestow a thought upon the perils of the sea. . . . Too late! She was hurled, helpless and struggling, to ruin and annihilation; and as she sank, engulfed, she carried with her the prestige of a race."[46]

Although history unmasked Dooner as a false prophet, in part, through the 1882 passage of the Chinese Exclusion Act and its decennial renewals, which barred entry to Chinese workers, his forecast of the "yellow peril's" eventuality was fulfilled in Japan's Pacific empire and attack on Pearl Harbor in 1941, according to some commentators.[47] Hailed as visionary by Clare Boothe, among others, who "discovered" it in the midst of World War II, was Homer Lea's *The Valor of Ignorance,* published in 1909.[48] With the U.S. eagle perched on its imperious aerie, Lea began, "the heights of universal history" afforded valuable insights into the rise and fall of states. "All kingdoms, empires, and nations that have existed on this earth have been born out of the womb of war," he declared, "and the delivery of them has occurred in the pain and labor of battle." If the United States was to attain greatness, it must, he insisted, "spread abroad over the earth the principles of its constitutions or the equity of its laws and the hope it extends to the betterment of the human race." And it must realize that expansion could be attained only through war preparedness and dominion over competing imperial powers in a world that was rapidly shrinking in size.[49]

Japan offered the most likely challenge to U.S. power in the Pacific, Lea proposed, and despite trade ties there existed an unbridgeable divide between them in race, religion and ethics, and social and political life. Still, though inevitable and years in the making, the war with Japan would break unexpectedly upon, as in Dooner's depiction, a naïve America, "strangely oblivious to the militant character of the Japanese, to the vast military and naval power in their hand, to the spirit of conquest in their bosoms, to their predetermined struggle with the Republic for sovereignty over the Pacific," that "vast Empire of Waters." Japan, in Lea's prognosis, with great cunning and duplicity, would invade and secure naval bases in the Philippines, Samoa, and Hawai'i and from there launch assaults on Alaska and the U.S. homeland. Without coastal defenses, the Pacific states would fall quickly to the invaders, and from their entrenched, elevated positions the Japanese army would easily repulse the American counterattack, weakened by the long, taxing march from the East over the plains and deserts of the vast interior. The defeated American forces would then scatter and plant

FIGURE 13. "The unsophisticated Mongol, imitating, ape-like, his fellow of this country, attains a monopoly of the cigar and laundry business, and smiles a cunning smile of triumph at his discomfited rival," reads the text to this cartoon: "The Coming Man: Allee samee 'Melican Man Monopoleeee," *The Wasp*, May 20, 1881.

"dissension throughout the Union, brood rebellions, class and sectional insurrections, until this heterogeneous Republic, in its principles, shall disintegrate, and . . . pay the toll of its vanity and its scorn."[50]

RACE WAR

Leading African Americans, including the oft-contending W. E. B. Du Bois and Booker T. Washington, held Japan, even imperial Japan, as the champion of the world's colored peoples during this period. In fact, that bond between African Americans and Asians was forged in the making of the U.S. empire in the Philippines, where African American troops, many of whom were veterans of the "Indian wars" in the Far West, formed easy identification with Filipinos—who were first called "Indians" by sixteenth-century Spaniards and "niggers" by twentieth-century U.S. imperial forces. Japan's defeat of Russia shattered the myth of white invincibility in modern warfare, rattling the global empire of whites and raising the hopes of African Americans and the colonized peoples of the tropical band.[51] Du Bois exulted: "The Russo-Japanese war has marked an epoch. The magic of the word 'white' is already broken. . . . The awakening of the yellow races is certain. That the awakening of the brown and black races will follow in time, no unprejudiced student of history can doubt." Booker T. Washington claimed, "In no other part of the world have the Japanese people a larger number of admirers and well-wishers than among the black people of the United States."[52]

Conceived of as a race war, that conflict was waged, on the one hand, for white supremacy throughout the world and, on the other, for decolonization and liberation from white rule and hegemony. In that context, even Japan's rapacious imperialism may be seen as an attempt to avoid or repulse white racism by emulating and exceeding white nationalism, empires, and ideologies, its racialisms against its enemies notwithstanding.[53] It was, after all, Europeans and Americans who launched voyages of exploration for trade routes to Asia that culminated with the empires of the modern era.[54] And it was they who mapped places and peoples and imposed their wills over them, inventing divisive and homogenizing classifications such as "white" and "colored" and "civilized" and "uncivilized" during imperialism's reign.

At the same time, Japan plied those waters and deployed race consciousness, posing as the liberator of colonized and oppressed peoples of the tropical band even as it displayed arrogance and practiced brutality toward them. Still, some African American leaders found irresistible Japan's siren call for a global alliance of the "darker races."[55] After all, Japan had loosened white supremacy's grip, first by defeating Russia and later by attacking the frontiers of European and U.S. empires, and its deeds and rhetoric, albeit two-faced, helped to agitate and accelerate the winds of change in the nationalist and decolonization movements that swept Asia and Africa after World War II.

Crucially, those notions of a race war were not limited to the realm of fantasy as in Pierton Dooner's "deductive history" but had material, palpable effects upon people's lives; some of the war's proponents possessed powers to actualize those ideas. At an outpost of the white homeland, strategists labored to shore up an "inner dike" against the "yellow peril's" rising tide. Sent to Hawai'i to investigate a sugar plantation strike, a federal commission raised the alarm in 1923: "It may be difficult for the home staying American citizen to visualize the spectre of alien domination which like a thunder cloud in the distance grows larger almost day by day, with the belief that when the infinite patience of this Asiatic Race [Japanese] has reached the point for action the cloud will break and America will wake up to the fact that it has developed within its Territory a race through whose solidarity and maintenance of Asiatic ideals will sweep everything American from the Islands."[56]

Earlier, an agent of the Bureau of Investigation, forerunner of the Federal Bureau of Investigation, had warned in a 1921 secret report, "It is the determined purpose of Japan to amalgamate the entire colored races of the world against the Nordic or white race, with Japan at the head of the coalition, for the purpose of wrestling away the supremacy of the white race and placing such supremacy in the colored peoples under the dominion of Japan."[57] Civilian and military planners braced for the impending conflict with Japan, accordingly, and despite preparations that failed to defend against the aerial attack of December 7, 1941, Hawai'i's rulers stood ready to exercise swift and efficient containment over the territory's Japanese from the moment of Pearl Harbor by mobilizing their decades-old plans for martial law, forced removals and detentions, and governing through fear.[58]

It was not, however, Japan that imported those Japanese workers to Hawai'i. The

FIGURE 14. The white man as burden: Chinese lift the British Prince of Wales during his visit to the Crown Colony of Hong Kong in the early 1920s. W. E. B. Du Bois observed "that dark and vast sea of human labor in China and India, the South Seas and all of Africa . . . that great majority of mankind, on whose bent and broken backs rest today the founding stones of modern history." W. E. Burghardt Du Bois, *Black Reconstruction in America* (New York: Russell and Russell, 1956), 15. Courtesy of Library of Congress.

Islands' rulers, the sugar oligarchy, had initially welcomed those migrant laborers, who now, with the onset of industrial strife and eventually war, were demonized as national security risks and the leading edge of Japan's radiating tidal wave of colored peoples that promised to engulf and overwhelm white supremacy worldwide. And it was Americans, not Asians, who annexed the Hawaiian kingdom and installed plantations of sugarcane and pineapple, with their capital, labor, and produce, as a nexus of a global order of "industrial communion" between the temperate and tropical zones, the United States and Hawai'i.

Unlike the earlier Caribbean plantations, those in the Pacific became widespread and grew in the 1860s after the American Civil War disrupted supplies from the South and created demands for sugar in the North and cotton for textile mills in Britain and elsewhere. The choice of Hawaiʻi by the United States for a favorable trade agreement, negotiated in 1876, as opposed to low-cost sugar producers in the Caribbean, was sweetened by the belief that the treaty would secure the Hawaiian Islands against other suitors. As the U.S. minister testified, Hawaiʻi, "if in possession of any European or Asiatic power, would be a standing menace to all the vital interests of the United States on our Pacific shores."[59] The kingdom's sugar plantations, tied to U.S. markets, would be a means by which to solidify U.S. imperial designs in the Pacific against the threat of imagined or real foreign advances along that frontier.

Typical of plantation systems, Hawaiʻi's sugar planters required imported laborers to tend their fields and factories, even as the indigenous population suffered catastrophic declines.[60] Although some among the kingdom's rulers wanted the immigration of "cognate races," including Asians and Pacific Islanders, to replenish the decimated Hawaiians, others preferred Europeans and Americans to "whiten" the complexion of the Islands' darker hue. Above all, the planters yearned for "servile laborers" or single men, without the expense of unproductive children, women, and the aged, who would work efficiently during the period of contract and would, when no longer gainfully employed, return home.[61]

Although that ideal worker was never fully realized, Hawaiʻi's planters, like other planters worldwide after slavery's end, found its equivalent in Asia, mainly in India and China, in the system of indentured or "coolie" labor. Predicting in 1853 a transition from slave to indentured labor in the Caribbean, William Thackeray recalled a story told him by an acquaintance of an "odd situation" that might alleviate the labor crisis brought on by the abolition of slavery in British colonies instituted in the 1830s. Gangs of Chinese "have been imported into Cuba," he reported, "who do the field-work so well, are healthy & orderly, & work at such a small price, that it is found that crops can be raised at a much less price than by the cumbrous & costly Slave machinery."[62]

Similarly, planters in Hawaiʻi enthused over Chinese and then Japanese contract or indentured laborers. An 1886 Bureau of Immigration report explained how plantations profited from Chinese workers: "Relying upon their work, at cheap rates, planters were enabled to purchase machinery, erect buildings, irrigate and drain on a large scale, all of which they would not have been able to do . . . had they not been able to count upon steady labor, at moderate rates, and for a stated term."[63] "We are in much need of them," wrote Robert Crichton Wyllie, Hawaiian foreign minister and master of Princeville plantation on the island of Kauaʻi. "I myself could take 500 for my own estates." Wyllie's letter, dated March 10, 1865, was addressed to Eugene M. Van Reed, an American businessman in Kanagawa, Japan. "Could any good agricultural laborers be obtained from Japan or its dependencies, to serve like the Chinese, under a contract for 6 or 8 years?" he inquired. Van Reed promised Wyllie's successor, "No better class of people for Laborers could be found than the Japanese race, so accustomed to raising Sugar, Rice, and Cotton, nor one so easily governed, they being peaceable, quiet, and of a pleasant disposition."[64]

The transformation of those pliant workers into the "yellow peril's" "swarming horde" was a process of conversion whereby presumably docile laborers proved intractable in the fields and resisted exploitation and physical and mental abuse, despite the impositions of a harsh and seemingly all-pervasive system of surveillance and punishment.[65] When no longer useful to their employers, "model" workers could undergo rebirth as "perils" and vanguards, in the case of peoples of color during much of the twentieth century, of a "race war." That "invasion," the migration of peoples and products from the tropical to the temperate band, was one of the fruits of European tropical empires. But other tropical fruits entranced empire builders like Andrew W. Preston, whose United Fruit Company came to dominate foreign lands, peoples, and governments, and who allegedly dreamed of the largely untapped tropics and its prodigious wealth.

4

Pineapple Diaspora

The romance of pineapples blossomed from a passion for a fruit with character, with attributes that distinguished it from other exotic, tropical fruits and objects of desire. That hankering, of course, was a human creation, which imputed natures to the plant and fruit, at first for Europeans as a rare and tasty reward of status and empire and, more recently, as a convenient and healthy product of industry and modernity. Human hands even shaped the fruit's so-called innate, native qualities.

The pineapple began its life as the terrestrial branch of the Bromeliaceae family, which humans selected, bred, cultivated, transported, marketed, and consumed. As was observed by J. L. Collins, a geneticist at the Pineapple Research Institute of Hawaii during the 1930s and 1940s, "The pineapple shares the distinction accorded to all the major food plants of the civilized world of having been selected, developed, and domesticated by peoples of prehistoric times and passed on to us through one or more earlier civilizations."[1]

Further, the pineapple and some of the most important crops in human history, corn, beans, potatoes, sweet potatoes, manioc, peanuts, tobacco, and cotton, are the gifts of the American tropics to the world.[2] And insofar as agriculture has allowed stable and large settlements, which enabled state formation and social complexities, America's migrating food crops—potatoes, corn, beans, and sweet potatoes—have had a profound impact upon the histories of Europe, Africa, and Asia. "Of all plant-movements," a study claimed, "the American plant-migration, although the most recent, is the most extensive, the most prominent, the most universal and the most momentous in the world's history. It therefore merits profound study in every detail. It has encompassed the globe in its entirety, made its influence felt everywhere, changed the surface of the earth and brought mankind together into closer bonds."[3]

NATIVE CULTURE

That the pineapple is indigenous to the "New World" is nearly certain,[4] but the place of the plant's nativity and the date of its domestication are indeterminate. On the basis of the geographic distribution of wild species, scientists believe that the pineapple probably originated in the general region where Argentina, Brazil, and Paraguay meet between the Paraná and Paraguay rivers. Here, at least three wild pineapple species grow in a variety of soils and tropical and subtropical climates, and it is likely that here native peoples domesticated the plant for its fruit. That location was found after a search, conducted between October 1938 and March 1939, for wild and cultivated forms of *Ananas,* the genus of the pineapple (from *nana,* the Tupí-Guaraní name for it, meaning "excellent fruit"). Though most still believe this place to be the pineapple's site of origin and spread, the definitive announcement contains a word of caution: "More botanizing and field study will be required before the area can be more exactly located."[5]

Botanists point to abundant evidence, in science and in reports by sixteenth-century European travelers, that natives found and propagated the seedless pineapple, which was a mutation of the wild, seedy varieties, some of which have so many seeds the size of rice grains that eating the fruit is extremely difficult and unpleasant. These peoples, they surmise, identified and selected several strains of pineapples, mainly for their fruit size, quality, and absence of seeds, then experimented with and made use of the plant's medicinal and fibrous properties and carried, transplanted, and distributed the pineapple wherever they traveled as traders and migrants.[6] One scientist noted that the pineapple's probable place of origin was home to the Tupí-Guaraní during "pre-Columbian times" and from that concluded that they, with the help of Caribs in northeastern Brazil, in the course of their migrations were responsible for the plant's dispersion along South America's Atlantic and Pacific coasts, the West Indies, and Central America.[7]

Tupí-Guaraní was one of the four major language families in sixteenth-century

FIGURE 15. Tupinamba family from Jean de Léry. A French cleric, Léry joined the 1555 French expedition to Brazil. In front of the "savages" is a pineapple. When ripe, he wrote, "you can smell them from far off; and as for the taste, it melts in your mouth, and it is naturally so sweet that we have no jams that surpass them; I think it is the finest fruit in America." From Jean de Léry, *Histoire d'un voyage fait en la terre du Brésil, autrement dite Amérique* (1580), published as *History of a Voyage to the Land of Brazil,* trans. Janet Whatley (Berkeley: University of California Press, 1990).

Brazil, and it encompassed a diversity of peoples and societies that changed over time. The primal proto-Tupí had words for "swidden" and "digging stick," indicating that some 4,000–5,000 years ago they practiced plant management. Their descendants, the proto-Tupí-Guaraní, ancestors of the Tupí-Guaraní, may have originated 1,800–2,000 years ago in the eastern foothills of the Bolivian Andes or on the Paraná and Paraguay rivers plateau area where humans perhaps first encountered the wild pineapple.[8] Archaeology reveals that between 3300 and 1800 B.C. the pineapple was grown along the lowland coast from Ecuador south to Peru.[9] The plant's discovery must have long predated its cultivation to allow for its selection and domestication or dependence upon humans for its survival and reproduction, and its dissemination and adoption also required time. Considering these factors and accepting the convention about its place of origin and its farmers, the proto-Tupí must have grown pineapples at least 4,000 years ago.

The pineapple's antiquity is not so remarkable; America's indigenous peoples, while often relying principally upon hunting and gathering for their nutritional and other needs, interacted intimately with and were highly skilled at identifying and exploiting a wide range of plants, including for food and medicine, and that botanical archive reached back to their first landfall on the continent.[10] As early as 8000–7500 B.C., Peruvians included in their diet two tubers, a legume, a fruit, and chili peppers, and by 5000–4000 B.C. inhabitants of northern Chile and Argentina ate several tubers, including the potato, along with corn, beans, squash, chili peppers, and guavas. About that time down to 1800 B.C., Andean hunter-gatherers added to that menu manioc, sweet potato, jícama, peanut, and avocado.[11] The pineapple was a contemporary of those dietary supplements and plants and environments with which the horticulturalists engaged.

The Bromeliaceae family has two branches, the terrestrial, those rooted in soil; and the epiphytes, which grow on other plants and objects but not as parasites. Pineapples belong to the terrestrial side of the family but share some of the attributes of their cousin epiphytes, such as the ability to store water in their leaves, making them drought resistant. Bromeliaceae are indigenous to the tropical and subtropical regions of America, which underwent dramatic climatic changes and witnessed a succession of plant, animal, and human activities contrary to the constancy and monotony imputed to the tropics by some Europeans. Humidity increased markedly between

12,000 and 9,000 years ago in the then-cool interior of central Brazil, and that increase was followed by a warming trend that led to a profusion of plant growth between 6000 and 2000 B.C. The *caatinga* ("white forest") in Tupí, an expanse of cactus and thorn trees, receded and gave way to the *cerrado,* a savanna with dense stands of trees. Those transformations affected cultures and ways of life, such as those of large animals that relied upon open plains and thus retreated before the expanding, thickening vegetation, and their human hunters, who turned to such other pursuits as fishing or migrated with the game.[12]

Given the paucity of evidence and the certainty of variation, we can only surmise at the world of the proto-Tupí or the people who perhaps gave us the pineapple. Most anthropologists believe that they were hunter-gatherers who relied upon game, fish, and wild and domesticated plants for their sustenance. They confronted a dazzling array of landscapes, waters, climates, and flora and fauna that challenged their ingenuity and creativity in provisioning for their needs and desires, organizing their social and political lives, and satisfying their intellectual and spiritual longings. By about 13,000 years ago, those Americans who passed through Panama's isthmus had occupied "every possible inhabitable environment throughout the continent—including the frigid straits and coniferous forests of the far south, the grassy pampas of modern-day Argentina, the rugged seasonally arid Brazilian uplands south of the Amazon River, the vast green sweep of the Amazon rain forest itself, and the awesome Andes mountain chain perched high over the western edge of the continent along its entire 7,700-kilometer length."[13]

Fieldwork in 1981 and 1982 among the Ka'apor, a Tupí-Guaraní people in the Amazonian state of Maranhão, Brazil, revealed that "humans and plants have complex relationships in which humans affect and nurture, as well as use, plants and entire plant communities." And though labor was generally shared, there were significant differences in how men and women allocated their time, men spending more of their day hunting, gardening, and fishing, women in food preparation, childcare, and manufacturing and repairing. Fairly sedentary and heavily reliant upon agriculture, the Ka'apor lived in a biologically rich environment with more than five thousand plant species, of which they recognized 179 edible, nondomesticated plant varieties; grew in gardens and swiddens (reclaimed forests that are not weeded) many of the more than one hundred species domesticated in the American tropics; distinguished between domesticated and nondomesticated plants; and altered and contributed to the Amazon's biodiversity.[14] That extensive inventory of plants, including their grouping into a classification system with specific names for their forms and parts, indicates their importance to Ka'apor women and men. If the Ka'apor division of labor by gender provides a clue for the distant past, it is equally probable that proto-Tupí women or men discovered and domesticated the pineapple. To be sure, that botanical body of knowledge was accumulated and transformed over time and space, indicating the necessity to historicize the passage of the pineapple and the people who created and carried it.[15]

In addition to the fruit's westward movement to the Pacific coastal lowlands, Tupí migrants, like the Ka'apor, may have transported the pineapple to the Amazon River basin and Atlantic seaboard, where they competed with the more established Aruak and, later, with the newly arrived Carib speakers. Although it is likely that the Caribs met the Tupí during the course of their southward expansion, the riverine Aruak may have originated in the Amazon Basin and from there spread northward in their canoes along the continent's littoral to Central America and across the sea to the Caribbean islands and Florida.[16] It is possible that in the course of those encounters among the Tupí, Aruak, and Carib the pineapple changed hands, and because of its portability the plant, in the form of tops, cuttings, or suckers, could have easily been stored in Aruak and Carib vessels.

Migration was a means of assuring productivity for those hunter-gatherers who tracked game and fish while cultivating crops such as corn, manioc, beans, sweet potatoes, squash, peanuts, and cotton in gardens and soils that were quickly exhausted and reclaimed by the forest. Movement also provided escape from invasions of people, intertribal feuds and warfare, and, much later with the advent of Europeans, epidemic diseases and enslavement.[17] Exemplary of the rapidity and spread of migration, a group of Tupí in the late sixteenth century migrated some 3,000 miles across Brazil from its Atlantic coast to the Andes and back to the Amazon's mouth over the course of just two generations.[18]

Besides its migrations as baggage, the pineapple may have been an object of exchange, following the trails of traders and their goods. If so, the pineapple's mobility would have hinged upon its economic value. But trade held wider significance than simply a system of transfer of scarce and desirable products. As was typical of both

MAP 5. America marked by boundaries of modern nation-states. Inset: region of the Ka'apor in eastern Amazonia. Adapted from map by Jeremiah Trinidad-Christensen.

continental and island Caribbean societie s of the fifteenth century, like their contemporaries from Europe, contact with distant peoples and the acquisition and display of
"luxury" goods were commonly reserved for the elite, indicative of their political and
religious prerogatives and powers. Exchanges may have established and solidified regional
bonds and hierarchies among disparate polities in diverse resource zones, from the cool,
mountainous highlands to the warm, humid low-lying valleys and plains, riverbanks and
seacoasts.[19] The pineapple, conveyed from great distances to the circum-Caribbean area,
may have been at first an exotic and rare fruit confined to the privileged class.

The Callinago, or Carib, Indians of the Lesser Antilles and other islanders traded
with and raided villages on the continent, from whence they originated and migrated
before 2000 B.C. Those settlers of the Lesser Antilles made pottery and practiced horticulture by at least 200 B.C., and they were in constant contact with peoples along
the east coast of Venezuela and the vicinity of the Orinoco delta.[20] The highway on
which they sailed their long canoes, the Caribbean, was an "ocean realm" and world,
"not as a group of circumscribed individual islands but as an archipelago, whose islands are linked by the ocean."[21] As a consequence of those initiatives, island cultures
are layered with different peoples and ways of life, as indicated by the materials they
left behind, including stone tools, ceramic ware, and mounds of shells, bones, and crab
claws. On the eve of their discovery of Europeans, the Callinago lived in small settlements usually built near freshwater rivers, and there, under headmen, extended families harvested the abundant provisions of the Caribbean Sea and employed *conuco*
agriculture, clearing and burning brush and heaping soil onto mounds to grow their
staple manioc (yucca, cassava), sweet potato, corn, beans, squash, peppers, peanuts,
tobacco, cotton, and pineapple, along with other fruits.[22]

EUROPEAN EXILE

The Callinago introduced the pineapple to Christopher Columbus and his men when
they happened upon the island the admiral called "Guadeloupe" in 1493 during his
second voyage to America. Known to America's indigenous peoples as a fruit that was
delicious, stimulated the appetite, and aided digestion, the pineapple had a flavor and

fragrance that, in the words of one account, "astonished and delighted" those alien visitors. Columbus, on his third expedition to America, found horticulturalists along Panama's east coast growing pineapples, bananas, and coconuts, among other crops, and he reported that the people there made a drink from the juice of the pineapple.[23]

Sixteenth-century European travelers took special note of their meetings with the *piña* (Spanish, pineapple) because of its unusual form, taste, and qualities, and they encountered it on several Caribbean islands, along the Central American coast from Panama to Mexico, and on the continent in Brazil, Guiana, and Colombia.[24] In his *De Orbo Novo,* Pietro Martir de Anghiera described the pineapple as an herb resembling a pine nut or artichoke grown in West Indian gardens and worthy of a king's table. King Ferdinand, one of Columbus's sponsors, ate a pineapple from Panama and gave it his highest praise. A 1519 foreigner to Brazil praised "a fruit resembling a pinecone, extremely sweet and savory, in fact the finest fruit in existence"; and Walter Raleigh, in his 1596 *Discoverie of the Large, Rich, and Bewtiful Empyre of Guiana,* amazed his readers with the opulence of the "New World," where there was "great abundance of pinas, the princesse of fruits, that grow under the sun, especially those of Guiana."[25] The Englishman, like Columbus, feminized and sexualized the land's temptations when, after sailing up the Orinoco in a quest for the elusive "Inca empire of El Dorado" (the golden man), he assured in language suggestive of rape, "Guiana is a country that hath yet her maidenhead, never sacked, turned, nor wrought."[26]

Diligent were the prolific and detailed firsthand accounts of America by Gonzalo Fernández de Oviedo y Valdés, sent by King Ferdinand of Spain in 1513 to oversee the production of gold and later appointed historiographer of the Indies. Part of his *Historia General y Natural de las Indias,* which appeared in 1535, contains the first published drawing of the pineapple. On Hispaniola, he recorded, grow "thistles" that produce a "pine, or better said artichoke, although because it looks like a pine cone the Christians call them pines, which they are not. This is one of the most beautiful fruits I have seen wherever I have been in the whole world." The pineapple, Oviedo declared, was matchless in "beauty of appearance, delicate fragrance, excellent flavour. So that of the five corporeal senses, the three which can be applied to fruits and even the fourth, that of touch, it shares these four things or senses excelling above all fruits, foods of the world wherein the diligence of man is occupied in agriculture. It has also another great quality," he added, "which is that without any trouble to the agriculturist it grows and sustains itself." Oviedo went on to describe the "prince of all fruits" and its varieties, both wild and domesticated forms, the plant and its leaves, its habitat and cultivation, and the fruit and its size, qualities, color, and usages, including as wine and medicine.[27]

Europeans took the pineapple—like their trans-Atlantic traffic of America's peoples—on board their ships as plunder and a prize of empire. Only royalty managed to taste the rare tropical fruit, other than the sailors who probably depended upon it (it lasted several weeks before rotting) to ward off scurvy. Before 1535, according to Oviedo, ships conveyed fruits and shoots to Spain, but most spoiled before arrival.[28] More successful were those fruits and plants that made shorter jumps from the West Indies to Virginia, where the pineapple thrived in 1614 until the winter's frost killed the plants, and from the West Indies to Bermuda in 1616–19, from whence the pineapple, along with the potato, sugarcane, and plantain, was reintroduced to Virginia in 1621.

During the first half of the sixteenth century, the Portuguese carried the pineapple from Brazil to Africa, Madagascar, India, and possibly China, and from China to the Philippines. In about 1660, the Dutch brought the plant back from Asia to their African colony at the Cape of Good Hope. An English embassy returned from China in 1657 with four pineapples, almost certainly preserved, as a gift for Oliver Cromwell, and four years later the "famous Queen pine" from Barbados was given to England's Charles II, an event that was perhaps documented in a painting by Dutch artist Hendrick Danckerts (fig. 17).[29] At a 1668 banquet hosted by Charles II for the French ambassador, the "rare fruit called the King-pine," native of the West Indies, made its appearance. "His Majesty," the diarist reported, "having cut it up, was pleased to give me a piece off his own plate to taste of," but it failed to live up to its reputation, perhaps because it was "impaired in coming so far."[30] That extravagant display may have been staged to show the French England's dominion in the contested West Indies.[31]

Besides acquiring them in the act of conquest, patrons at home in Amsterdam grew pineapples imported from Dutch colonies in Java, Surinam, and Curaçao, and a Flemish merchant, Le Cour, kept plants in his hothouse near Leyden.[32] The English, French, and Germans followed the Dutch, who apparently were the first to succeed in bringing the pineapple to fruit in the cold north, around 1690.[33] About that same time, the pineapple was introduced to England as a botanical specimen, a marooned native of the

FIGURE 16. "I do not suppose that there is in the whole world any other of so exquisite and lovely appearance," exclaimed an enraptured Gonzalo Fernández de Oviedo y Valdés. "My pen and my words cannot depict such exceptional qualities, nor appropriately blazon this fruit so as to particularize the case fully and satisfactorily, without the brush or the sketch." Drawing in Oviedo y Valdés, *La historia general de las Indias* (1535), f. 37v; courtesy of the Huntington Library, San Marino, California. Quotations from J. L. Collins, *The Pineapple* (London: Leonard Hill, 1960), 10, 11.

tropics, which was finally coaxed into producing fruit by Henry Tellende, gardener for Matthew Decker of Richmond, Surrey, using heated glasshouses in about 1720. Theodorus Netscher celebrated the achievement in a painting (fig. 18) that bore the inscription: "To the eternal memory of Matthew Decker . . . and of Theodore Netscher. . . . Here the pineapple, worthy of a royal feast, grew at their expense at Richmond."³⁴

Pineapple growing sparked intense competition among the wealthy of Europe, emulating national rivalries, each estate trying to exceed the other in producing a superior fruit. German Otto von Münchhausen built large hothouses in his gardens near Hanover exclusively for pineapple cultivation. A grateful Gottfried Wilhelm Leibniz

FIGURE 17. John Rose, the royal gardener, presenting a pineapple to England's King Charles II, c. 1670s. Painting attributed to Dutch artist Hendrick Danckerts. Source: Ham House, Surrey, United Kingdom.

praised Münchhausen in 1714 for giving to Germans a fruit that "all the travelers in the world would not have given us" through a "method of multiplying them, so that we may perhaps have them one day as plentiful, of our own growth, as the oranges of Portugal, though," he tempered his enthusiasm, "there will, in all appearances, be some deficiency in the taste." An English visitor to the king's table in Hanover reported, contrary to Leibniz's caveat, that Münchhausen's pineapples were "perfectly delicious." Those natives of Brazil, she explained, were brought to "such perfection" by "stoves" that lengthened the summer "as long as they please, giving to every plant the degree of heat it would receive from the sun in its native soil."[35]

FIGURE 18. Theodorus Netscher's painting of a pineapple grown in Sir Matthew Decker's garden (1720), Richmond, Surrey, England. Source: Fitzwilliam Museum, University of Cambridge, United Kingdom.

Growing techniques were kept secret during the initial contest for personal and national stature, but gradually gardeners shared their cultural techniques and deliberated the attributes of different varieties in horticultural journals. Scottish writer John Claudius Loudon introduced the *Gardener's Magazine* in 1826, making accessible plant varieties, gardening tips, and layouts of model gardens to the expanding middle class, and his book *The Different Modes of Cultivating the Pineapple from Its Introduction into Europe to the Late Improvements of T. A. Knight, Esq.* (1822) detailed pineapple culture, including the construction of hothouses in which to grow pineapples

FIGURE 19. Erected in 1761 by John Murray, the fourth Earl of Dunmore and later governor of colonial Virginia, the Dunmore Pineapple is one of the architectural landmarks of Scotland. Located in Dunmore Park near Airth, this garden retreat and hothouse sheltered pineapple plants warmed by a furnace-driven heating system that circulated hot air through a gap between the double walls. Source: Gistimages/Alamy.

FIGURE 20. Botanical illustrations from live plants, like this "Queen pine" from the garden of Belgium's Prince Leopold, as published in Charles McIntosh, *The Practical Gardener* (1828–29), were key elements in the work of scientific identification. Realism prevails in this rendering of the luscious, yellow fruit surrounded and capped by leaves with sharp thorns.

and other exotic, tropical plants. In that work, Loudon reported that shipping had improved and cultivation in the West Indies had increased, such that large consignments of pineapples from Bermuda reached England in six weeks, which "has lessened their price and rendered them common." That development led to the decline of pineapple production in European hothouses and the rise of plantation production in tropical America during the first half of the nineteenth century. Still, Europeans

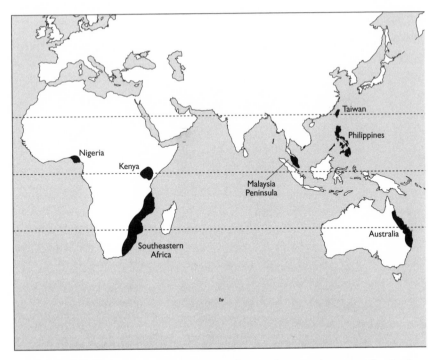

MAP 6. Centers of world pineapple cultivation about mid–twentieth century, showing its clustering within the tropical and former European colonial world. After J. L. Collins, *The Pineapple* (London: Leonard Hill, 1960), 76. Adapted from map by Jeremiah Trinidad-Christensen.

continued to experiment with new varieties of pineapples in hothouses of tropical climates created by natives of the temperate zone, despite the abating of the tropical fever that had invaded and swept the large estates of European landowners.[36]

The wealthy and powerful classes of America and Europe alike, in exhibits of privilege, reserved for themselves commodities—rare, expensive, and desirable—from far-off places. The pineapple was not simply a delicious fruit; the "princess of all fruits" came to symbolize the tropics, the Orient in opulence, leisure, a terrestrial paradise. Its possession accordingly meant the attainment of social standing and its trappings. The fruit's top, or "crown," though manifestly regal, terminated in sharp thorns along

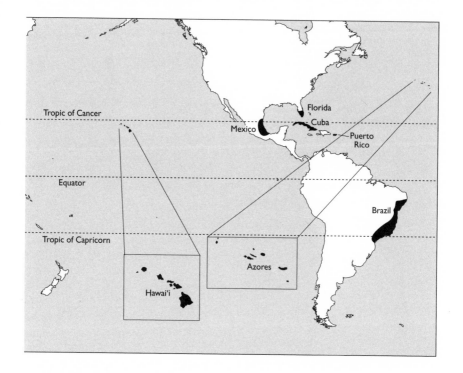

with the plant's spiny leaves, and its skin covered, with scales, or "eyes," was rough to the touch. So while Oviedo was fulsome in his praise of the fruit's "lovely appearance" and "delicate" taste, he noticed the plant's "very sharp thorny thistle with long prickly leaves," disclosing that it was at core, like America's natives and flora and fauna, "very wild."[37] Paradise, Europe's Orient, was indeed both civil and savage.

Though the fruit was still uncommon and expensive, it graced the tables of the wealthy class of Europe and was oftentimes the centerpiece and apex of a mound of fruits of various kinds—the source, like the Orinoco, of a copious stream. Its demand for banquets and dinners was such that a single fruit was sometimes rented for several parties and moved from one table to another.[38] Accordingly, the portable pineapple "carried with it a subtle implication of an elite social standing, because it had long been the exclusive prerogative of wealthy and educated people." The pineapple as a symbol of ostentation migrated from Europe to the United States as decoration on silver and ceramic dinnerware, appearing on American tables by the eighteenth century.[39]

CAYENNE CROSSINGS

The pineapple, as a food and object of social and economic value, circulated the globe on the currents of European commerce and colonization. Exemplary of that pattern is the sojourn of the Cayenne variety of pineapple, *Ananas comosus,* which originated in the American tropics, was removed to Europe, and from the temperate zone was re-distributed to tropical colonies for plantation production.[40] Perhaps first cultivated by the Maipure in the upper reaches of the Orinoco River, in the vicinity of Paradise as held by Christopher Columbus and Walter Raleigh, the Cayenne pineapple has since supplied profits and income for many thousands and fed millions more worldwide.

In 1819, France dispatched an expedition to its colonies in America and the Pacific to collect seeds and plants for botanical gardens in Paris and Versailles. At Cayenne, the capital of French Guiana, expedition member Samuel Perrottet reported finding a new variety of pineapple, which he named *Bromelia mai-pouri.* "This new species of *Ananas* was procured from Cayenne," he wrote; "there were five plants as I have said, deposited in the garden of new varieties at Versailles. The mai-pouri does not have spiny leaves like its relatives; its fruits, of a very delicate flavor, weigh on an average 10 kilograms (20 pounds) and are very fine."[41] "Mai-pouri" was probably the variety's name in French Guiana and indicated its place of origin, Maipures, a village at the junction of the Triparro and Orinoco rivers, or its cultivators, the Maipure, who had long occupied that area. The plant migrated, carried by indigenous travelers and traders, to the coast and what later became French Guiana.

By 1820, five transplanted Cayenne, as told by the expedition's botanist Perrot-tet, grew in the gardens of Versailles, as specimens and spectacles from the far reaches of the French empire. Those additions to the Versailles Royal Kitchen Garden joined others of their migrant kin in the "pinery," consisting of "a vast number of pits, succession and fruiting houses, the whole of which are heated with hot water; it contains

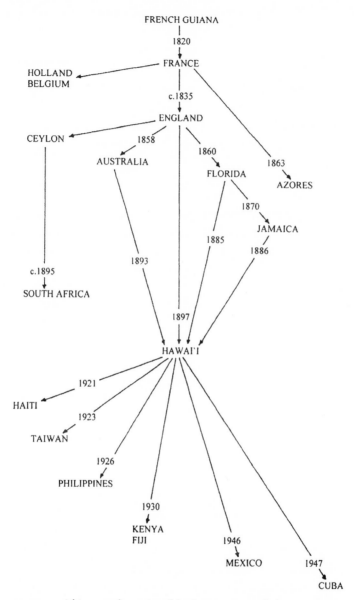

FIGURE 21. A history and mapping of the Cayenne pineapple diaspora, from its origins in America to its exile in Europe and remigrations from Hawai'i. Source: Gary Y. Okihiro, after a diagram in J. L. Collins, "Notes on the Origin, History, and Genetic Nature of the Cayenne Pineapple," *Pacific Science* 5, no. 1 (January 1951): 5.

about 2000 plants . . . in all about 40 varieties." The pineapples thrived in the "pinery," a tropical zone in France, such that by 1841 more than a thousand plants of various kinds bore fruit each year, and from its original five the Cayenne variety had multiplied to produce three hundred fruits annually.[42] Three growers in France offered the stock for sale, and the Cayenne, because of its smooth leaves and fruit with "excellent flavor and weight on the average from 9 to 12 pounds," as an advertisement boasted, soon became a favorite among European gardeners.[43]

From those nurseries in France, the Cayenne crossed the channel to England, and from that center it was transported to British colonies Australia by 1858, and Jamaica, in 1870 via Florida. From Florida in 1885 and Jamaica the following year, the Cayenne migrated to Hawai'i, and new infusions of that variety from Australia entered Hawai'i during the 1890s.[44] By the late nineteenth century, Hawai'i was a pineapple hub, receiving varieties, including the Cayenne, from the tropics and Algeria, Australia, the Bahamas, Florida, Jamaica, Mexico, Puerto Rico, Samoa, Singapore, and Trinidad. In turn, during the first half of the twentieth century, Hawai'i was the redistribution core of the Cayenne diaspora to Africa and other islands of the Pacific and back to its original American home.[45]

These remarkable movements of the Cayenne pineapple across islands, oceans, and continents track the rise and course of European empires impelled, in large measure, by desires. Captives of those wants and initiatives, Indians and their material culture, including the pineapple, were spoils of imperial designs, annexations of conspicuous wealth and power that encircled the globe. Possessions they were of conquest and the ruling class, and as such they were procured, collected, and displayed prominently in gardens and museums and in the intimacy of dining and drawing rooms. Even their environment yielded to glass houses that conjured the tropics in the temperate homeland by harnessing the sun's heat year-round. But it was expensive, that mastery over nature, especially when improved navigation and speedier ships drew the tropical band closer to Europe. So, instead, settler colonies and tropical plantations, tried first in the mild Mediterranean, disciplined native and imported labor and exploited the resources of near and distant lands and seas. The career of the pineapple and its transits from the tropical to temperate worlds and back bear ample witness to those acts of generation and dispersion.

5

Hawaiian Mission

About the time the proto-Tupí began cultivating the pineapple in America, proto-Polynesians left their Southeast Asian homeland for anticipated opportunities to their east. Those skilled sailors and horticulturalists settled New Guinea's northeast coast to form "Lapita culture" around 1500 B.C., and groups of them continued sailing eastward against ocean currents and winds until, around 1000 B.C., they reached what we now know as western Polynesia. Centered on Tonga and Samoa, the migrants indigenized and evolved to become "Polynesians." Yet, like the natives of the Caribbean islands, they remained a seafaring people, and their island world was large, providing them the resources, physical and metaphysical, of land, sea, and sky. Expeditions of Polynesian men, women, and children with their plants and livestock on board larger, more advanced canoes designed for greater distances reached the Marquesas around 100 B.C., Easter Island off America's west coast around A.D. 300, and New Zealand about 500 years later.

Several Polynesian companies violated tradition when they pointed their canoes northward, contrary to the successful east-west migration trail, across unfamiliar seas. Guided by the daily arc of heavenly bodies, the swirls of ocean currents, the color of the water and its biotic communities, and the skies with their clouds and airborne animal life, Polynesians made landfall in the islands they called Hawaiʻi as early as A.D. 300, and they were followed by others of their kin from the Marquesas and, later, the Society Islands around 1100. Successive groups brought with them new ideas of religion, economy, and government along with crops and animals, and they, like their island-hopping counterparts in the Caribbean, layered the earlier custodians of the land and sea and their social formations.

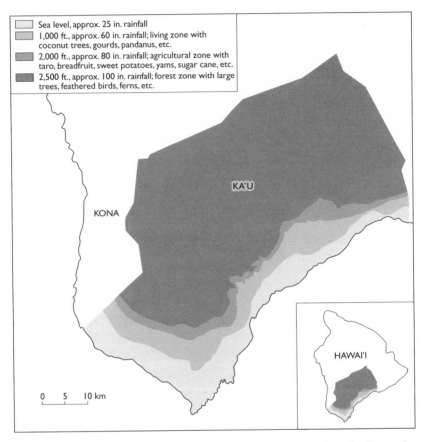

MAP 7. Kaʻū district on the island of Hawaiʻi, showing its land divisions from the shore to the uplands with their rising elevations, rainfall, and products. Adapted from map by Jeremiah Trinidad-Christensen.

NATIVE LAND

Amidst the abundance of the newfound islands, the voyaging Polynesians created the Hawaiian people, engaging water, earth, and sky filled with life, seen and unseen, and kinfolk all. "The Hawaiian people," a historian explained, "are the living descendants

FIGURE 22. Genesis and kinship are established in the taro, whose shoot used in propagation is called the 'ohā; the 'ohana (family), offshoots of the taro, and ali'i (royalty) are called "taro planted in the land—a source of life." Gary Y. Okihiro photograph, 2008. Quote from E. S. Craighill Handy and Elizabeth Green Handy, *Native Planters in Old Hawaii*, Bulletin 233 (Honolulu: Bishop Museum Press, 1972), 76.

of Papa, the earth mother and Wakea, the sky father. They also trace their origins through Kane of the living waters found in streams and springs; Lono of the winter rains and the life force for agricultural crops; Kanaloa of the deep foundation of the earth, the ocean and its currents and winds; Ku of the thunder, war, fishing and planting; Pele of the volcano; and thousands of deities of the forest, the ocean, the winds, the rains and the various other elements of nature." In fact, she noted, "The Hawaiian people are a part of nature and nature is a part of them."[1]

By the late eighteenth century, the Hawaiian Islands supported the largest, most densely populated society of all the Polynesian islands, indicating the productivity of its lands and waters. Horticulture was the mainstay of the economy, and the primary unit of land was the *ahupua'a*, a narrow, wedge-shaped slice of land, marked off by an *ahu* (altar), that started from an apex in the highlands and widened out toward the coastline. With this arrangement, Hawaiian producers had access to all of the land's resources, including the tall trees and colorful birds of the higher elevations, the low-land fields for their planted crops, and the ocean's bounties along its shores, reefs, and the deep beyond.

The windward, better-watered side of islands allowed intensive cultivation of taro in terraced fields and wetlands supplied with water from springs, rains, and irrigation ditches, while on the leeward side farmers grew dry-land crops such as the sweet potato. Although migrating birds may have carried seeds to the Hawaiian archipelago long before Polynesians arrived there, the seafaring humans probably were the bearers of food crops and domestic animals to the new lands, including varieties of taro, sweet potatoes, yams, gourds, bananas, sugarcane, coconuts, and breadfruit along with dogs, chickens, and pigs.[2] Those introductions by humans simply added to the Islands' biotic communities brought there by oceanic and atmospheric currents and through the agency of birds, fishes, and other life forms.

FOREIGN DESIRES

At dawn they were spotted, on January 18, 1778—British captain James Cook's tall ships, which were on their way to America's Pacific Northwest to find the fabled passage that supposedly connected Atlantic with Pacific and therewith a shorter route from Europe to Asia. The strangers wanted fresh water, pigs, chickens, bananas, taro, and sweet potatoes, and some of the sailors desired the company of women. In the exchange, the foreigners left bits of iron, goats, melons, pumpkins, onion seeds, and possibly venereal disease.

At first a mere "refreshment" station, Hawai'i's islands soon became a favored destination, a purveyor of commerce as a provisioner of ships involved in the China trade and, later, in whaling and a supplier of sandalwood trees for the Chinese market. The islanders, mainly men but also a few women, served as sailors and servants on board those ships and as soldiers and hunters in the Pacific Northwest for the furs desired by Chinese buyers. And when Yankee whalers crisscrossed the Pacific in search of their diminishing quarry, Hawaiians were recruited for those crews depleted by afflictions and desertions.

Foreigners overwhelmed Hawaiian initiatives mainly because of diseases, which decimated the population, and desires that seduced Hawaiians into an economic system favorable to its creators. "If a big wave comes in," prophesied Davida Malo in 1837 of the European flood, "large and unfamiliar fishes will come from the dark ocean, and when they see the small fishes of the shallows they will eat them up." A Christian convert and graduate of mission schools, Malo still warned against those "large and unfamiliar fishes" and their appetites while witnessing the steep and swift decline of the Hawaiian population. "The white man's ships have arrived with clever men from the big countries," he observed. "They know our people are few in number and our country is small, they will devour us."[3] Europeans called Hawaiians "Indians," and like America's natives they suffered horrific losses after contact with whites. Variously estimated at 250,000 to 800,000 in 1778, the Hawaiian population had plummeted by more than 50 percent by about the time of Malo's premonition of his people's dispossession by foreigners.[4]

Although the chiefs suffered population losses comparable to the commoners, they were able at first to broker their privileges to benefit their class. Kamehameha, the archipelago's first king, instituted a royal monopoly over foreign trade and distributed trade rights of prized commodities like sandalwood among a few select chiefs to ensure their loyalty. Many among the Hawaiian elite were avid clients of imported luxuries, including the king, who was described as "a man of powerful physique, agile, supple, fearless, and skilled in all the warlike and peaceful exercises suitable for an alii [chief] ... a strong mind ... well filled with the accumulated learning of his race and capable of thinking clearly and effectively. He exhibited much curiosity in regard to new things and new ideas and showed good judgment in adopting those calculated to promote his interests." He was a conspicuous consumer and "something of a dandy, clad in 'a colored shirt, velveteen britches, red waistcoat, large military shoes and worsted socks, a black silk handkerchief around his neck.'"[5]

Few commoners enjoyed those choices under the new economic order, and instead they felt the heavy burden of increasing taxation upon their labors aimed at supplying the foreign trade. Similar to the earlier demands of military service and war

mobilization exacted by the chiefs, felling sandalwood in the interior forests and haul-
ing it to waiting ships at the coast diverted efforts away from food and craft production to an extractive commerce, which depleted both Hawaiian resources and labor
and undermined local industries. And as in times of war, famine and displacement
accompanied the new, colonial-like economy, hastening the reduction of Hawaiian
numbers and self-sufficiency.[6] Wage labor offered a measure of escape from chiefly
demands and an eroding social order, although it too threatened life and limb. Some
women worked as prostitutes to obtain goods and cash, and so many men joined the
crews of trade ships that during the 1840s the kingdom levied a poll tax on ship captains to compensate for the labor lost to the Hawaiian economy. Still, in 1850, about
12 percent of all working-age Hawaiian men departed to voyage like their ancestors,
but on foreign ships and to strange landings outside their ocean.[7]

For mercantilists in the United States, the seductions of the China trade were immense, helping to produce ostensibly the first millionaire, Elias Derby of Salem, Massachusetts, from profits that soared, in some instances, to 700 percent.[8] Typical of Yankee traders was the Boston firm of Bryant & Sturgis, which instructed its captain of
the brig *Ann,* bound for Hawai'i in 1818, to add to the ship's crew twenty-one or
twenty-two "stout Islanders" to help sail to and gather fur pelts in the Pacific Northwest. From there the captain was to return the Hawaiians to the Islands and pay them
off in goods, "sell the King any articles of your cargo on advantageous terms" for a full
load of sandalwood, and sell the furs and sandalwood in China for Chinese goods.
American traders, wrote a historian, "descended upon the islands in a swarm, bringing with them everything from pins, scissors, clothing, and kitchen utensils to car-
riages, billiard tables, house frames, and sailing ships," and they created needs and stimulated desires, "doing their utmost to keep the speculating spirit at fever heat among
the Hawaiian chiefs."[9]

The merchants were so successful that the kingdom's foreign debt quickly soared
to $200,000 according to an 1826 estimate, and when Hawaiian clients neared bank-
ruptcy the U.S. government commissioned two warships to investigate and enforce
its traders' claims. Thus compelled, the chiefs imposed the earliest written tax law,
dated December 27, 1826, which required every able-bodied man to deliver a specified
load of sandalwood or its cash equivalent of $4 (Spanish) and every woman, a mat,

FIGURE 23. Lahaina, Maui, called "one of the breathing holes of Hell" by a missionary, was a popular port for New England whalers. The steepled building on a hill beyond the town is Lahainaluna Seminary, where Davida Malo and other Hawaiians trained for Christian service. Detail from *Whaling Voyage around the World* (1848) by Benjamin Russell and Caleb Pierce Purrington. Courtesy of New Bedford Whaling Museum.

barkcloth, or $1 (Spanish) in cash.[10] As a consequence of Hawaiian participation in that marketplace, the once dense stands of sandalwood trees, like the native peoples, diminished, and the trade in sandalwood, begun in 1790 and so associated with the Islands that the Chinese knew Hawai'i as the "Sandalwood Mountains," came to an end by the 1830s.[11]

U.S. commerce with the Hawaiian Islands, an 1829 account summarized, involved on average 125 ships each year, some carrying sandalwood and furs, others in transit between America and Asia, and still others in pursuit of whales, and their total value amounted to over $5 million. Between 1852 and 1859, on average, 484 whaling ships visited Hawaiian ports, and the domestic produce supplied to all ships exceeded $150,000 annually. At first managed by Hawaiian chiefs, the provisioning of ships fell increasingly under the control of foreign trading firms, which built storehouses and retail stores mainly in Honolulu to mediate the transfer of goods. James Hunnewell, an agent for Bryant & Sturgis, opened a business in 1826 that evolved into C. Brewer & Company, one of the five firms that came to dominate Hawai'i through the twentieth century. The others of this "Big Five," except Alexander & Baldwin, which got its start in sugar, had similar beginnings in mercantilism. Castle & Cooke, begun by American missionaries, imported and sold merchandise; Theo. H. Davies & Company was originally the British commercial house Starkey, Janion & Company; and a German merchant, H. Hackfeld, opened a store in Honolulu that became the final of the Big Five, American Factors.[12]

PACIFIC POSSESSIONS

From its first mapping onto European grids by James Cook in 1778, Hawai'i was steadily drawn into a global network of empire and capitalism. European penetration of the tropics, stimulated by desires for Asia's products as well as by Europe's vision of the Orient, was the principal transgression that accounted for Hawai'i's position in the new world order. Spain's "discovery" of America and its "Indians" was a step toward Asia, and indeed its settler colonies extracted from the indigenous peoples the gold and silver necessary for the trans-Pacific traffic that made Spain the wealthiest nation on earth by 1600. From 1565 to 1815, connecting two of Spain's Pacific possessions, galleons carried American precious metals, sugar, tobacco, grains, and cowhides from Mexico's Acapulco to purchase Asia's spices, teas, porcelain, ivory, silks, and cotton in the busy markets of Manila, trade center of the Philippine Islands, named to honor Philip II. An estimated one-third of the silver mined in Mexico and Peru went into this trade, which also involved the mission Indians of New Spain, who were encouraged by their spiritual fathers to hunt for the sea otter pelts valued in China. The Manila galleon of 1783 departed Acapulco with more than seven hundred such pelts in its hold, and between 1786 and 1790 California's missions collected from their communicants offerings of sea otter skins worth over $3 million.[13]

Spain's Pacific dominance lasted through the sixteenth and seventeenth centuries and was eclipsed by principally the British in the next century. As early as 1575, English raiders had preyed on Spanish ships and ports in the Caribbean, and Francis Drake plundered several of Spain's Pacific coast settlements up to northern California, crossed the Pacific to Asia and Africa, and returned to England with his *Golden Hind* laden with Spanish bullion. Increasingly after 1600, European nations ignored

the exclusive claims of Spain and Portugal to the tropical world as had been delimited by a 1493 papal bull that, in complete disregard of native peoples, gave to those
two countries divine rights to "pagan" lands.

The English, Dutch, French, and Russians staked their claims in the Pacific by occupying this oceanic basin and naming and classifying the nature of its seas, islands,
and peoples. The British Admiralty and Royal Society, accordingly, underwrote James
Cook's three expeditions during the 1760s and '70s for the advancement of both science and empire. Cook's instructions were "to determine [the Pacific's] strategic rele-
vance to British imperial interests" by charting waters, lands, and skies, cataloguing
plants, animals, and peoples, and speculating upon trade possibilities. It was on his third
and final Pacific expedition that Cook put Hawai'i onto European maps and initiated
a scramble for furs when his men found, like the Spaniards before them, that the Pacific
Northwest's sea otter pelts were exceedingly profitable in the China market.[14]

Soon after the War of Independence, the United States joined its former master
in pursuit of Pacific plunder when in 1784 financier Robert Morris and other in-
vestors dispatched the *Empress of China* from New York's harbor with its cargo of
New England ginseng for Canton. That beginning and the anticipation of greater
returns led the United States to expand its frontier by circumnavigating the continent and depositing outposts along what became its western littoral. The nation's two
frontiers, one by land and the other by sea, would eventually meet in the Far West;
the nation's destiny lay across the expansive "marine continent" called the Pacific, in
the Far East. Although the venture capitalists reaped a mere 20 percent return on
their investment in the *Empress*'s trial run, its example and report stirred dreams of
fantastic fortunes along with a contempt for perceived Chinese arrogance and self-
sufficiency.[15] An independent, ancient China apparently offended some of the elite
of the independent, infantile United States.

China would soon be brought to its knees by the British, who stopped the drain
on the gold and silver that paid for their tea addiction by using agents of the British
East India Company to push opium among the Chinese as early as the 1790s. Opium,
derived from the company's holdings in India, was the means by which the British
addressed their balance of trade problem. The cash-poor United States relied upon
poached natural resources, including ginseng, furs, and sandalwood, to exercise its

youthful imperial ambitions, and within the span of a mere two decades "American maritime frontiersmen," a historian wrote, "like their continental counterparts in confronting the North American heartland, were becoming adjusted to the grandeur, expanse, challenge, and prospects of their awesome milieu." Their enthusiasm for commerce was such that by 1800 more than one hundred Yankee ships made port in Canton, and the Americans were second only to the British in the volume of that trade.[16] To secure and improve their hold, U.S. visionaries conceived of settlements in Pacific harbors as havens for and suppliers of ships, and as military bases from which to impose dominion over an "American lake."

MERCY MISSION

New England missionaries, too, were empire builders, commissioned to recreate Hawaiian society in their own image of Christianity and capitalism. In a turn from inland American Indian ministrations to overseas burdens and prospects, the American Board of Commissioners for Foreign Missions outfitted and launched an expedition to Hawai'i in October 1819. "Your mission is a mission of mercy," the board instructed its first company, "and your work is to be wholly a labor of love." This "arduous enterprise" was "a great and difficult work" of uplift. "You are to aim at nothing short of covering those islands with fruitful fields and pleasant dwellings, and schools and churches; of raising up the whole people to an elevated state of Christian civilization."[17] Churches and schools were the means by which to work a transformation in Hawaiian thinking, while "fruitful fields," a key element in the mission schools' curriculum, helped to channel Hawaiian industry. Conversions of both body and mind sought to bury the "old" and give birth to the "new."

Missionaries hastened the advent and kingdom of capitalism by inducing changes in the pattern of land tenure and system of labor. "We deem it proper for members of this mission to devote a portion of their time to instructing the natives into the best method of cultivating their lands, and of raising flocks and herds, and of turning the various products of the country to the best advantage," an 1838 meeting of missionaries on Maui held. "The missionary should endeavor to call forth ingenuity, enterprise, and patient industry, and give scope for enlarged plans for profitable exertion."[18] Unstated but motivating that effort was the indigence of the Hawaiian mission and the need to sustain itself in the light of meager funding from home made even tighter after new stringencies were imposed in 1837.

A response from both the board and missionaries in the field during those hard times and attempts at income generation was to accuse each other of having strayed from the original spirit of self-denial, seeking instead "personal aggrandizement, luxury and ease."[19] In fact, the call for Hawaiian efforts in cultivation was a way to tend to mission as well as to personal expenses in the field.[20] Two years earlier, in 1836, the Hawaiian mission proposed an ambitious initiative for a Christian colony, a spiritual kingdom on earth in which commercial plantations and manufacturing industries would employ the masses in useful labor and supply revenues to the government (and mission).[21] The gospel of "profitable exertion," as envisioned in the doctrine of Christian capitalism, failed in the abstract although it ultimately triumphed in reality.

Caught in the web of foreign trade and mounting deficits, along with occasional visits and threats by armed ships and marines to force concessions, the king and chiefs relied increasingly upon the settlers, including missionaries, who advised the kingdom and served its offices. Thousands of Hawaiians converted to Christianity, especially after the "great awakening" of 1836–39,[22] and many more received training in mission schools, making the Islands ripe for the harvest. The commerce in sandalwood gave way to whaling and resident mercantile houses, and Hawaiians of all classes were swept along those powerful currents, which impelled them to produce for the market, away from self-sufficiency and toward employment or servitude on foreign ships.

Those developments led to greater demands on the part of the growing settler community to consolidate and secure its economic, political, and spiritual gains. The settlers found their opportunity when nine-year-old Kamehameha III ascended the throne in 1824. The child depended upon his Christian mentors, and as a young man, in 1840, he signed a constitution that installed a constitutional monarchy with executive, legislative, and judicial branches of government in which, from 1842 to 1880, foreigners occupied twenty-eight of the thirty-four cabinet seats and 28 percent of the legislature, though they totaled only 7 percent of the population.[23]

CATERPILLAR HORDES

Not all of the Hawaiian people, including those of the chiefly class, were content with the alienations of sovereignty. An eyewitness to the momentous events unfolding before him, Samuel Kamakau, historian and statesman, testified that "the Puritan missionaries instructed us that the law was the established word which determined the rights of kings, chiefs, and commoners." Likening the chiefs and commoners to "a luxuriant tree," Kamakau charged that the king acted like a *kaunaoʻa* vine, which "draws from the tree sustenance for itself, drying it up, draining it dry, and destroying it."[24] In addition, "the chiefs objected to placing the new constitution over the kingdom, seeing that little by little the chiefs would lose their dignity and become no more than commoners." But those chiefs who might have stood fast against a ruler forgetful of his responsibilities to the people were dead, he lamented, and the king and his foreign advisors ruled the day.[25]

The commoners, the *makaʻāinana,* in a petition dated July 22, 1845, urged the king not to depart from "the old good ways of our ancestors" to follow "the new good ways." For, they warned, there are men "living right among us who will devastate the land like the hordes of caterpillars the fields; they hide themselves among us until the time comes, then they will be on the side of their own land where their ancestors were born." Those scheming men, the petitioners held, with the king's assent, reversed the positions of privilege in a way that Hawaiians were now debased and foreigners exalted. "The Hawaiian people will be trodden under foot by the foreigners," they predicted. Further, "the dollar is become the government for the commoner and for the destitute. It will become a dish of relish and the foreign agents will suck it up." So although the aliens demean the natives as "stupid, ignorant, and good-for-nothing" and cause the chiefs to depend upon them and "cast off their own race," the *makaʻāinana* are not blind to the injustice. "The laws of those governments will not do for our government. Those are good laws for them, our laws are for us and are good laws for us, which we have made for ourselves. We are not slaves to serve them. When they talk in their clever way, we know very well what is right and what is wrong."[26]

In 1845, Kamehameha III moved his capital from Lahaina to Honolulu, signaling a shift away from the influence of the Maui chiefs to the settler business enclave and locus of foreign dominance.[27] In that same year, the government enacted laws that allowed foreigners to become naturalized citizens of the kingdom and acquire title to lands. Fifty-two Hawaiians signed a petition in protest: "Do not sell the land to new foreigners from foreign countries," they pleaded with the king. "We have heard of this sale of land to foreigners. There is aroused within us love and reluctance to lose the land, with love for the chiefs, and the children, and everything upon the land. We believe we will soon end as homeless people." Naturalized foreigners, they warned, would claim the status of "true Hawaiians" and evict the native peoples from their land. The new economic order would lead to that, the use of land for revenue, for wealth, for "worldly goods." "Listen to the voice of wisdom announcing to you in this petition," they advised. Preserve the land "as it is very valuable." Conserve the people, their independent government, and their king, and thereby save them from "the foreigners."[28]

LAND GRAB

The land that nurtures, ʻāina, was the crux of the matter. Before their discovery of Europeans, Hawaiians tended Papa, mother earth, as a sacred trust and in veneration of the ancestors who were embodied in nature. The living land that breathed and grew with Pele's activities could never be conceived of as owned, as private property, and even after unification under Kamehameha I land was held in common by all of his subjects despite his conquest. As much was attested in the Hawaiian saying "Born was the Land, born were the Chiefs, born were the common people," indicating the union of land, chiefs, and people,[29] and the 1840 constitution explained that, although "all the land from one end of the Islands to the other" fell under Kamehameha, "it was not his own private property. It belonged to the chiefs and people in common, of whom Kamehameha I was the head, and had the management of the landed property." At the same time, the king and chiefs parceled out the land to their people, and with those assignments came rights to the land and its products and duties to the chiefs in the form of taxes on goods and labor.[30]

Foreigners, who like commoners depended upon the chiefs for land, negotiated their interests, unlike commoners, in treaties with Britain (1836) and France (1837) by which the Hawaiian government pledged protection of foreign-held lands as a species of property. The king tried to rationalize the erosion of sovereignty: "We renounce the right of dispossessing them [foreigners] at pleasure. We lay no claims whatever to any property of theirs, either growing, or erected on the soil. That is theirs, exclusively. We simply claim the soil itself, but do not claim that even that should be restored, though from the old we have never had the smallest idea of alienating any portion of our land. But if the soil be not restored, then we claim a reasonable rent."[31] These concessions, together with an 1839 declaration and the 1840 constitution that, although citing historical precedent for communal landholding under Kamehameha I, articulated rights to land and property and lay the foundations for the rush on "worldly goods" feared by many Hawaiians. Agricultural production supplied foreign ships and later distant lands, provided revenues to the government, attracted investments, and employed Hawaiians in useful labor for the glory of God and mammon. In those pursuits and land uses, settlers were distinctively unlike the *makaʻāinana*.

The Board of Commissioners to Quiet Land Titles, created in 1845 to ascertain and settle land claims, announced the Great Mahele, or land division, three years later.[32] The division reserved about 24 percent of the kingdom's lands for the crown, 39 percent for 245 chiefs, and 37 percent for the government. Under that division, commoners could make land claims against chiefly lands, and those were finalized by an 1850 act that allotted to 8,205 *makaʻāinana* 28,600 acres, or less than 1 percent of the land. Although all adult males were eligible, only 29 percent filed validated claims, leaving 71 percent landless. The legislation stipulated that claims could be made only on lands "really cultivated," a condition inviting interpretation and explaining why so many of them were disallowed. Those were made permanent by a subsequent act that similarly required proof in four years of "real" cultivation.[33] To further the dispossession, accompanying the land division for Hawaiians was another 1850 act that granted foreign settlers the right to buy, own, and sell land in Hawaiʻi.

Settlers, whose land desires had the sanction of law and who had access to more capital than their Hawaiian hosts, were quick to seize that opening and entitlement. In 1850 the Hawaiian mission petitioned and received from the government 560 acres for each missionary, and between 1850 and 1860 the government auctioned off some of its land holdings, 64 percent of which went to foreigners, the rest to Hawaiians.[34] The coup, a legal study pointed out, was a triumph of Western imperialism "without the usual bothersome wars and costly colonial administration," and the doctrine of Christian capitalism had won the day. In January 1850, former missionary Richard Armstrong boasted gleefully to his brother: "The government has lately granted fee simple titles to all natives, for the land they have lived on and occupied. This gives the final blow to the old feudal system and makes this a nation of freeholders. It is a point for which I have long contended and finally on my own motion it was carried by the King and Council."[35]

In Hawai'i since 1832, Armstrong quit the missions to head the Ministry of Public Instruction in 1847. This "father of American education" in the kingdom placed manual labor in the schools' curriculum, agriculture for boys and homemaking for girls, in the belief that "early habits of industry will supply their wants, make their homes comfortable and remove the temptation to wander about and commit crime in order to get money or fine dress." Further and of no small consequence was his hope that the profits earned from student labor would help to defray the costs of their education.[36] Armstrong knew that the privatization of land would compel the Hawaiian chiefs to hire their former tenants as wage laborers and "sell their waste lands of which they have an abundance."[37] The net result of those timely interventions was Hawaiian alienation from the *'āina,* from the source of their mana, or power, from sovereignty, caused by the seizure, occupation, and transformation of Hawai'i's island world by foreigners.

The kingdom sold more than 600,000 acres at an average price of 92 cents per acre, with most of the land falling into the hands of foreigners, who now held extensive tracts. For instance, before 1864 a mere 213 foreigners bought more than 320,000 acres; and a year later a solitary entrepreneur bought the entire island of Ni'ihau, more than 61,000 acres. Not only did these land transfers have the blessing of law, many had the stench of fraud. Courts upheld questionable transactions, and crown lands, though inalienable since 1865, were leased for pennies per acre. In 1880, Claus Spreckels, the California sugar baron, bought all of the crown lands inherited by a descendant of Kamehameha I, and when the sale was voided, he convinced the legislature to give him 24,000 acres of high-quality sugar lands in return for relinquishing his rights to his illegal purchase. Marriages were another means by which white men acquired Hawaiian property, including the more than 370,000 acres of the estate of Bernice Pauahi Bishop, wife of Charles R. Bishop and the last descendant of Kamehameha I. Upon her death in 1884, five white men administered her charitable trust, and a majority of the justices of the supreme court had the prerogative to name their successors. Accordingly, even though the Bishop estate was devoted to the welfare of Hawaiians, control of its holdings, representing nearly 10 percent of the Islands, passed from the people's trusteeship.[38]

MISSIONARY CONVERSION

Men's fortunes flourished in this "virgin" land. Some of the earliest white men pursued lives of ease by marrying Hawaiian women, like the dozen or so European men in the kingdom during the 1790s who reportedly were "not employed at anything since the wives that each has maintain them."[39] Charles R. Bishop, on his way to Oregon, became the king's collector of customs within three years, served as minister of foreign affairs, founded the kingdom's first bank, and married Princess Bernice Pauahi, the largest landholder in Hawai'i. A Welshman, Theophilus Harris Davies, gained control of a bankrupt merchandising house during the sugar boom, financed twenty-two sugar plantations, and in 1875 began the Honolulu Iron Works Company, which manufactured the machinery required on plantations and in sugar mills. A stranded sailor from Massachusetts, Benjamin F. Dillingham, wed a daughter of missionaries and sired a financial dynasty of extensive land holdings, a railroad, and a ship terminal with wharves and warehouses. A carpenter on Maui, James Campbell, amassed tens of thousands of acres of land on O'ahu and Hawai'i. Among foreigners long-standing in the kingdom, missionary families like the Alexanders, Baldwins, Castles, Cookes, Hitchcocks, Rices, and Wilcoxes were zealous for capitalism, and they were well represented on the boards of directors of all the major firms in the Islands.[40]

That pattern of land tenure in which a few assumed possession of extensive tracts, typical of plantation societies worldwide, enabled the plantation economy that dominated Island life for almost a century. The first sugar plantation set the example for those to follow. Ladd and Company, an American mercantile firm in Honolulu, signed a lease with Kamehameha III and Kauaʻi's governor on July 29, 1835, for about a thousand acres for a term of fifty years at an annual rent of $300. In addition, the agreement stipulated that the plantation would retain the ability to hire native workers who would be exempt from chiefly taxes upon their labor and products but for whom the company would pay the king and governor 25 cents a month for each laborer and compensate the workers with "satisfactory wages."[41] That contract, initiated by an American mercantile firm for large acreage for pennies per acre over a long term, was precedent setting, as was the disruption of the ties between subsistence horticulturalists and their land and chiefs for wage labor.[42]

The following year, William Hooper of Boston, a Ladd and Company partner "scarcely twenty-six years old, without experience in agriculture, the mechanical trades, or sugar production," sailed for Kauaʻi to begin his company's business venture with a keen sense of mission. The Koloa sugar plantation, he recorded in his voluminous diary, was established "for the purpose of breaking up the system . . . or in other words to serve as an entering wedge . . . [to] upset the whole system."[43] Hooper, like Richard Armstrong, believed that plantation capitalism would, indeed, deliver the final blow to the "old" and breathe life into the "new." In truth, men like Hooper and Armstrong and a host of New England missionaries, while ostensibly working for Hawaiian regeneration, found a new faith and were themselves reborn, experiencing conversion in those tropic isles.

6

Tropical Plantation

"Strange indeed were the hard thoughts of the missionary!" observed Samuel Kamakau, likening their devotions to a feast. "So they [the missionaries] girded up their loins, sharpened their knives, and chose which part of the fish they would take, one the side piece, another the belly, one the eyes, another white meat, and another red meat. So they chose as they pleased. When the last man of them had come they were treated like chiefs; lands were parceled out to them. . . . Yet they found fault. Now you want to close the door of heaven to the Hawaiians," Kamakau objected. "You want the honors of the throne for yourselves because you sit at ease as ministers upon your large land."[1]

A visitor disagreed with the Hawaiian scholar's assessment. Not all missionaries "made themselves notorious by neglecting their spiritual duties for temporal affairs," he claimed. Instead, many saw themselves as the defenders of Hawaiians against foreign lusts and fraud.[2] And yet their object to raise the people up "to an elevated state of Christian civilization," as was their instruction, mirrored their belief in their superiority, moral and otherwise.

PRODUCING PLANTATIONS

The ministry's paternalism, when examined in the light of world history, represented a notable turn in fortunes. Europeans were not always the civilizers of benighted peoples but were latecomers to and impoverished supplicants in a global system of trade during the "Asian Age," according to a provocative study featuring Indian, Chinese, and Islamic peoples from as early as 1400 to 1750.[3] Europe's seeking after Asia offered proof of its desire to emulate and subdue its estimable other. Fantastic Persia and India astonished the ancient Greeks with their wealth, abundance, and exotic, tropical products; commerce with India after Alexander's conquests brought ivory, spices, and peacocks and parrots into Greece and, later, Rome. The stranglehold of middlemen on overland routes prompted European initiatives on the oceans, such as Roman trade with Asia that centered on Alexandria and the Red Sea. From India came women slaves, skins, furs, and hides from Tibet, Siberia, and Turkestan; wool from Kashmir; ivory, pearls, and tortoise shells from the Indian Ocean; precious stones from India; and silks, bronzes, and pottery from distant China. Such was the demand for those pleasures during the first century A.D. that "moralists" complained about "the drain of precious metals from Rome to pay for them," and Pliny fussed "over the large sums of money spent on luxuries for the ladies."[4]

The rise of Islamic civilization after about A.D. 700 brought the tropics into closer communion with the Mediterranean basin. Rice, taro, coconuts, oranges, lemons, limes, sugarcane, plantains, bananas, and mangoes came from India and Southeast Asia, carried by the Arab expansion across Africa to Spain along with tropical crops from sub-Saharan Africa.[5] Making their way northward from Africa, many of those crops would become important to the economies of Cyprus, Sicily, Spain, and Portugal; in turn, Spain and Portugal, with their ventures into the tropical world, spread crops both within the tropical band and between the tropical and temperate zones. The Mediterranean basin, accordingly, served as a fertile breeding ground for transmissions and innovations, as was exemplified in the patterns of land tenure and labor that typified plantation economies in the cultivation of sugarcane.

Indigenous to New Guinea, sugarcane accompanied human migrations eastward, with, for example, peoples who would later become Hawaiians, and westward to India, where records describe the production of *gur* about 500 B.C. and sugarcane growing much earlier. A fourth-century B.C. Sanskrit administrative manual records five sugar varieties, including *khanda,* which became the English "candy." The Chinese knew of sugarcane cultivation to their south about 200 B.C., and they grew and manufactured sugar some 400 years later. From northern India, where sugar technology was preeminent, the crop moved to Persia around A.D. 600, and from there farther west to the shores of the Mediterranean, where, for nearly a thousand years, producers in Palestine, Egypt, Sicily, Spain, Morocco, Cyprus, and Crete supplied most of the sugar for the area.[6]

Conquest of Muslim-held lands during the crusades of the late eleventh century enabled a form of agriculture different from the feudal system of northern Europe. On Cyprus, the main center of sugar production for Europe from the thirteenth to fifteenth century, lords who received land grants as rewards employed local and foreign enslaved laborers without the constraints of kinship, custom, or law. Managing their estates as businesses, families and the clergy employed workers, including slaves from Arabia and Syria and migrant laborers from Palestine, built massive irrigation systems necessary for cane cultivation, imported from Europe boiling kettles and equipment for sugar refining, and exported their granulated sugar to Europe.[7]

Although the plantation system reached maturity in the Atlantic world, that Mediterranean beginning bore aspects of its fruition, involving conquered lands, an exploitable workforce often coerced and imported because of local population declines, a capitalist agricultural enterprise, a single crop destined for foreign markets, and control of the system by foreigners or foreign nations.[8] According to one historian, plantations were "agricultural units in tropical and semitropical areas, typically engaged in monoculture or dual cropping, export oriented, under foreign ownership and management, and dependent on a sizable, servile, and low-paid labor force."[9] Moreover, established by and for Europeans, plantations with their spread were tied "to the wider world economic community in very precise ways." They were planted in tropical soil with capital from the temperate band and tended mainly by colored labor to produce tropical exports for white planters. Plantations, thus, were "only one part of a much wider world economic system consisting of a set of relations which meet at a metropolitan and industrial centre far removed from the plantations."[10]

The plantation, with the expansion of Europeans, would become a familiar fixture in the tropical world. With major centers of sugarcane growing in Morocco and the south of Spain, its leap to islands off the West African coast, many uninhabited, was simply a matter of Portugal's occupation and colonization of them. Madeira, the Canary Islands, and São Tomé became key sugar producers for their masters. Madeiran sugar reached European tables after 1450 and within fifty years was found as far east as Constantinople. Sugar from São Tomé, farther south and near the equator, arrived in Europe during the 1490s; sugar from Brazil began arriving in the early 1500s. The new American plantations, blessed with abundant and cheap land and labor and an ideal climate, soon eclipsed the Mediterranean sugar producers as Europe's main suppliers. Besides the lower prices of Atlantic sugar, political instability in the region and a failure to keep pace with technological innovations accounted for the final collapse of the Mediterranean sugar industry in the sixteenth century.[11]

That westward shift, from the Mediterranean to the Atlantic, signaled not only the rise of the Atlantic plantation complex but also an Atlantic world instigated by European nations bordering that ocean, their extractions of gold and silver from America for trade with Asia, their enslavement of the indigenous peoples of America and Africa, and their colonial implantations in America. Those movements eventuated in Europe's ascent and Asia's decline, the underdevelopment of Africa and America, and the creation of the modern world-system. Needless to say, a host of other factors were involved in these transformations, such as the emergence of mercantile capitalism and city-states, the rise of nations, the development of cities and industries and the correlated advance of agricultural production, along with decays in Asia before the arrival of European ships in the fifteenth century.

Still, insofar as human encounters and dependencies constituted a system, all of its constituent parts and their articulations were relevant to the operation of the whole. Simply put, the temperate and tropical zones were connected by circulations of people and goods, even as there were concentrations of capital and labor. And the development of a region and people had its counterpart in the underdevelopment of another region and people. Tropical plantations embodied those theoretical propositions.

KING SUGAR

Drawn into capitalism's expanding net, Hawaiian destinies shifted from a north-south orientation as the apex of the Polynesian cultural triangle to an east-west bearing as the crossroads of the Pacific in the commercial traffic between America and Asia. And with the increase and influence of settlers from the United States, the Hawaiian Islands moved closer to the Northeast, from whence came the whalers, traders, and missionaries, but also to the Pacific coast, where Hawaiians assisted in the harvest of furs destined for China and worked and settled the land, interacting with whites and the continent's native peoples. The steady U.S. westward advance by sea and over land from the eastern seaboard, as if manifestly destined, converged on that other shore, where Hawaiian producers contended with rapidly growing white settlements of consumers and producers.

The California gold rush exhausted Honolulu supplies of pickaxes and shovels, clothing, boots and shoes, and flour in the fall of 1848, and "clamorous purchasers" eagerly carried off coffee, sugar, fruits, and vegetables, creating shortages in the Islands and false expectations of rising future demands. Irish potatoes that sold for $2 a barrel in Honolulu commanded $27 in San Francisco, and accordingly production and exports experienced a boom, reaching a peak in 1850 though slumping a mere two years later because of cheaper crops available in Oregon and, later, California.[12] "The extension of the territory and government of the United States to the borders of the Pacific," the newly formed Royal Hawaiian Agricultural Society exulted in 1850, "the wonderful discoveries in California and the consequent creation of the mighty state on the western front of the American continent, has as it were, with the wand of a magician, drawn this little group [Hawai'i] into the very focus of civilization and prosperity."[13]

Most of the Islands' agricultural products desired by whites were, with few exceptions, European introductions, including potatoes, corn, coffee, pumpkins, melons, onions, cabbage, string beans, tomatoes, limes, and oranges, along with cattle, sheep, goats, and different varieties of chickens, dogs, and pigs. Exceptional were the once large and magnificent sandalwood stands, natives to the archipelago, and some of the food crops and animals the Hawaiians brought with them from Polynesia. Sugarcane later acquired commercial value, and pineapples may have been carried to Hawai'i by Polynesians or been introduced there by Europeans.[14] Called *hala kahiki* ("foreign pandanus") by Hawaiians, the pineapple, as that name suggests, may have been an immigrant, although some Hawaiians on Maui and Hawai'i, where whites first reported seeing it growing shortly after Cook's expedition, claim the fruit as native to the Islands. If they are correct, the pineapple, like the sweet potato, may have been conveyed from America to Polynesia by natives of America or Polynesia long before the arrival of Europeans in the Pacific.[15]

For the most part, foreigners, whether as visitors or settlers, sought to channel production toward their wants, and explorers and scientists such as James Cook and their expeditions not only collected and classified the flora and fauna of the lands they visited but also spread food crops and livestock for their benefit.[16] Citrus combated scurvy; tubers, wheat, corn, and meat stored well and provisioned ships' galleys; and certain tropical fruits, like pineapples, maintained their freshness longer than many other fruits. So, although whites may have designed peaceful gardens and sprawling ranches at home and abroad ostensibly for civilizing inferior races and plant and animal species, they also dreamed of plantations, colonies, and markets in an expansive, acquisitive, and invasive empire of wealth.

Sugar in Hawai'i, once grown only on small plots for local consumption, rose in significance with its export and entry into the world market. Although its attendant difficulties frustrated its founder, William Hooper, the Koloa plantation steadily expanded its fields, improved its mill and refinement process, produced a better grade of sugar, and increased its output. Koloa's example resulted in the demise of individual cultivators and small mills, replaced by more efficient, larger operations of land, labor, and capital investment. By 1846 there were eleven sugar plantations, two run by Chinese who were pioneer sugar producers,[17] and sugar exports advanced quickly from 4 tons in 1836 to 180 tons in 1840 and nearly 300 tons in 1847. As was anticipated by Hawai'i's planters, California's and Oregon's growing populations energized the cultivation of cane and the transformation of labor and land tenure in the Islands. Sugar interests prospered and land values rose until high tariffs, competition from other sugar producers in the Pacific, and a glutted U.S. market led to the depressions of 1851 and 1852, which bankrupted many planters. The setback was, however, short-lived.

Cut off from the South's raw materials during the Civil War, the North presented a ready market for higher-priced imported sugar and cotton. For instance, the price of a pound of sugar in the North rose from 6.95 cents during the 1850s to 10.55 cents in 1863 and reached its peak a year later at 17.19 cents. The inflated prices of that distant market spurred increases in acres under cultivation in Hawai'i, from 2,750 in 1853 to nearly 22,000 in 1882, and during the period 1860–70 sugar exports swelled from 702 to 9,392 tons.[18] The 1876 Treaty of Reciprocity with the United States continued that upward trend in Hawaiian sugar exports and production, the number of plantations growing from twenty in 1875 to sixty-three in 1880, cane acreage increasing some one thousand percent, and sugar exports, which totaled 25 million pounds in 1875, ballooning to 250 million in 1890. As early as 1879, just three years after the treaty, the *Pacific Commercial Advertiser* proclaimed that trade figures showed "quite plainly that sugar is King."[19]

These developments illustrated the ties that bound the tropics with the temperate zone not as "complementary trade regions," in the words of geographer Ellen Churchill Semple, but as a species of economic imperialism exemplified by the "banana empire" of Andrew W. Preston and his United Fruit Company in the Caribbean. And like the United Fruit Company, which was a business conglomerate that assumed the prerogatives and functions of government, the Big Five that controlled Hawaiian sugar during this period also ruled as an oligarchy over the Islands' economy and government. Further, following its tentacles, Hawai'i's colonial economy was closely tied to the financiers and refineries in the San Francisco Bay area, and ultimately to markets in the Northeast. The competitors and sometimes partners of Hawaiian plantations included other tropical plantations of sugar and pineapples in the Philippines, China, and the Caribbean.[20]

ISLAND OLIGARCHY

The Big Five firm C. Brewer & Company had its start as an outfitter of whale ships and turned to financing and developing four sugar plantations in 1866. Castle & Cooke added to its merchandise business by issuing loans to plantations and by 1870 had become the agent and manager of four sugar plantations.[21] As one study recounted, "The factors who grew to dominance after . . . [the U.S. Civil War] were agents of the plantations for obtaining tools and supplies, floating loans, and selling sugar. The early spread of the factor system is explained readily by the advantage of specialization." While plantation managers worried over deploying labor and technology, the merchants in Honolulu, especially after the end of the whaling era, supplied the necessary capital and equipment for plantations, managed transportation, and supervised foreign exchanges and markets. "The vital lack of capital, and the superior access which the commercial houses had to capital, gave them a strategic advantage in their dealings with plantations," the study explained. Overexpansion and economic downturns produced a situation "in which plantations were faced with either bankruptcy or the necessity of turning stock over to their factors," leading "to the complete domination of the sugar industry by the largest factors."[22]

By 1910 the Big Five controlled 75 percent of Hawai'i's sugar, and 96 percent by 1933, and they managed every aspect of the business. From banking to transportation, a 1906 federal investigation revealed, the Big Five lived off the plantations, "taking their profits in lean years as well as in fat ones"; "the compensation that capital receive[d] from the industry [was] greater than appears on the surface." That concentration of economic power was kept within a circle called "the family compact," in which missionary families and their relations through marriage predominated. At least one missionary descendant served on every Big Five board during the early twentieth century, and through a system of interlocking directorates the family compact was observed. An oligarchy, the Big Five wielded both economic and political power, installing governments at will and serving in their offices. "There is a government in this Territory which is centralized to an extent unknown in the United States, and probably almost as much centralized as it was in France under Louis XIV," boasted the attorney general, a relative of the governor, Sanford B. Dole, son of missionaries from Maine, and one of five members of that family to serve in key positions in the Dole administration.[23]

That territory, in fact, testified to the powers of the missionaries and foreigners who, in Kamakau's words, lusted after "the honors of the throne" while sitting "at ease as ministers upon your large land." To be sure, Hawaiians participated in and resisted those erosions at every turn,[24] while suffering devastating population losses and assaults against their religion, culture, and economy that drained away the kingdom's lifeblood. Despite direct opposition to and attempts to ignore the foreign invasion, there was no escape from the necessity of confronting its realities.

NATIONAL SOVEREIGNTY

King Kalākaua (1874–91) illustrated some of the tough dilemmas faced by Hawaiians. Elected king by the privy council and legislature as specified by the constitution and thus beholden to them, Kalākaua called upon his benefactors, Britain and the United States, to land troops to quell the riots that greeted the news of his selection. Hawaiian nationalists had supported Queen Emma, widow of the late king, and they feared the loss of their nation to foreigners. Kalākaua, his mana gifted by whites, was the only ruler whose selection Hawaiians demonstrably rejected. In January 1875, less than a year after his elevation, Kalākaua proposed the treaty of trade reciprocity with the United States that gave an immense boost to the sugar and business interests in the Islands and gave them more influence in the kingdom. In addition to that commercial tie, which was extremely favorable to the United States, the treaty forbade the Hawaiian kingdom to give economic or political preferences to any nation other than the United States, rendering it, in effect, a U.S. dependency.[25]

The sugar boom, coupled with the Hawaiian population decrease, stirred the worldwide recruitment of labor for the plantations and search for a "cognate and friendly race" to secure the demographic and political future of the kingdom.[26] Those two needs revealed Kalākaua's quandary. Migrant workers increased the economic and political powers of the white planters, which the king sought to counter with a robust, "native" population. Fresh infusions of a race similar to Hawaiians, such as the Japanese, the king thought, would through assimilation and intermarriage augment the numbers and powers of the "native" as opposed to "foreign" peoples. As the king's envoy to Japan, John Kapena, told the Japanese emperor: "Hawaii holds out her loving hand and heart to Japan and desires that your people may come and cast in their lots with ours and repeople our Island home, . . . [that they] may blend with ours and produce a new and vigorous nation making our land the garden spot of the eastern Pacific, as your beautiful and glorious country is the western." That international alliance of Oceania's nations, the king believed, might ensure his kingdom's sovereignty and assist in ridding the Pacific of Western imperialism.[27]

Accordingly, on his 1881 visit to Japan, Kalākaua urged the actualization of an earlier proposed plan of Japanese migration to Hawai'i and the creation of a Union and Federation of Asiatic Nations and Sovereigns sealed, he confided to the emperor, by the marriage of his niece, Princess Kaiulani, to a Japanese prince.[28] Japanese migration to Hawai'i would resume and swell a few years after the king's visit,[29] but mounting a challenge of colored peoples to white global supremacy, the threatened "yellow peril" and "race war" hoped for by African Americans W. E. B. Du Bois and Booker T. Washington, among others, would have to wait for another day.

Japan, under the duress of its modernization campaign and claims to equal footing with Western nations, especially the United States, reluctantly declined Kalākaua's "profound and farseeing views." Mutsuhito, the Meiji emperor, in a letter to the king dated January 22, 1882, explained his decision. "The Oriental nations, including my country, have long been in a state of decline and decay," he began, "and we cannot hope to be strong and powerful unless . . . [we] restore to us all attributes of a nation." The union proposed by the Hawaiian king was a step in that direction of independence from "those powerful nations of Europe and America," and that move was "a pressing necessity for the Eastern nations, and in so doing depend their lives." But the task of rebuilding Japan, as opposed to crafting international alliances, was the more urgent of the two. "However," Mutsuhito confessed, "I ardently hope that such Union may be realized at some future day, and keeping it constantly in my mind I never fail," for that alliance would not only secure "the fortune of Japan and Hawaii, but also of whole Asia."[30]

Clearly, despite an international standing that took him to Japan, the United States, and Europe, Hawai'i's king was vulnerable to political machinations at home. Preserving the nation's independence, as the Japanese emperor reminded him, required great skill and courage in balancing and choosing between competing demands. Whites, comprising missionary, commercial, and foreign interests that at times clashed, were a powerful and persistent tribe. "Hawaii for the Hawaiians," a slogan adopted by both Kalākaua and Queen Emma, articulated the other side of the equation—one that was equally varied but at core united around the identity of the nation and its subjects.[31] Native desires for sovereignty had to be accounted for, particularly because Hawaiians were the majority of the voting population and nationalists held the legislature from 1874 to 1887, throughout most of Kalākaua's eventful reign.

Kalākaua's appointment of the foreign yet Hawaiian nationalist Walter Murray Gibson as prime minister in 1882 signaled a shift away from alien, white power, and he ignited a Hawaiian cultural revival through the Hale Naua Society to give prominence to Hawaiian learning, a Board of Genealogers to establish the legitimacy of the chiefs over foreign claimants and usurpers, and a Board of Health empowered to license native healers.[32] Through renewals from "cognate races" and the pursuit of indigenous knowledges, Kalākaua strived to achieve his aim, Hoʻoulu Lahui (Increase the Nation).[33]

When the Reciprocity Treaty, with its terms renewed annually by an oft-reluctant Congress, expired in 1885, the anxiety level of those who profited from the treaty rose even as Hawaiian nationalism flourished. In 1886 the white elite formed the Hawaiian League, a group, they claimed, dedicated to "good government" "by all necessary means" under the label "Hawaiian" and not "foreign." Despite the fact that most whites refused to give up their U.S. or British nationality for citizenship of "this very small kingdom," their ample investments compelled political dominance to ensure those bounties.

An assertion of nativity without its attendant loyalties and duties was the objective of the "Hawaiian" League when in 1887 it forced upon Kalākaua the Bayonet Constitution, which reduced even further the king's powers and enfranchised qualified men of "American or European descent." That racialized, gendered provision did not require a renunciation of previous citizenship or a declaration of allegiance to the Hawaiian nation, though it rendered foreign men equal to native men in the vote. The league had engineered, through the Bayonet Constitution, a coup d'état. At a protest rally held two months after that display of white supremacy, a Hawaiian leader exhorted the crowd, "Now tell me, have any of you been endangered by this new constitution?" To which the people replied, "We are greatly oppressed."[34]

Demonstrations against the 1887 constitution and reform government culminated with an insurrection in 1889, led by Robert W. Wilcox, in which seven nationalists were killed, a dozen wounded, and about seventy arrested. First charged with treason and then conspiracy, Wilcox was subsequently acquitted by an all-Hawaiian jury.[35] On January 20, 1891, after speaking to his people on an Edison recording machine, Kalākaua lapsed into a coma and died in San Francisco on his way to negotiate a renewal of reciprocity. His sister and regent, Liliʻuokalani, was named queen. A known opponent of the Bayonet Constitution, the queen was supported by thousands of registered voters, who urged her to enact a "new constitution for our country and our people."[36] Those signatures were "the voice of the people," Liliʻuokalani wrote, and "no true Hawaiian chief would have done other than to promise a consideration of their wishes."[37]

ALOHA ʻOE

While the queen drafted a new constitution to restore the powers of the monarchy, members of the white elite secretly worked with John L. Stevens, the U.S. minister to the kingdom, for annexation by the United States and formed the Annexation League in 1892. Lorrin Thurston, a league founder, went to Washington, D.C., to gain support for annexation, and there he was told that President Benjamin Harrison assured that, when "you come to Washington with an annexation proposition, you will find an exceedingly sympathetic administration here." Thus encouraged, when on January 14, 1893, the queen announced her intention to abrogate the 1887 constitution, the Annexation League saw a "splendid opportunity to get rid of the old regime, and [make] strong demands for annexation, or any kind of stable government under the supervision of the United States." Members resolved to declare "a Provisional Government with a view to annexation to the United States," gathered arms to enforce their revolution, and consulted with Stevens, who stood ready to land U.S. marines to protect American lives and property, ostensibly, but mainly to recognize and defend the provisional government.[38]

The next day, Liliʻuokalani conceded defeat and agreed to abide by the Bayonet Constitution, but league members pursued their "splendid opportunity" and charged the queen with "revolutionary acts" by which "the public safety is menaced and lives and property are in peril." Unable to save themselves from the "general alarm and terror" that they had instigated, the usurpers asked an expectant Stevens "for the protection of the United States forces." That petition was signed by thirteen whites, six of whom were Hawaiian citizens, five were U.S. citizens, one was British, and one, German.[39]

On January 16, 1893, U.S. marines landed, in the words of President Grover Cleveland in a report to Congress, "for the purpose of supporting the provisional government," and not to protect American lives and property. A select group of the league met the following day and appointed Sanford B. Dole president of the new government while Stevens rejected the queen's appeal for support. Asserting that they acted for the common good, those pretenders to the throne declared an end to the Hawaiian kingdom and imposed a provisional government "until terms of union with the

United States of America have been negotiated and agreed upon." Liliʻuokalani refused to acknowledge that illegitimate takeover and, knowing that the U.S. minister
"aided and abetted their unlawful movements and caused United States troops to be
landed for that purpose," in her words, surrendered instead, under protest, to "the superior force of the United States of America."[40]

The provisional government proclaimed, on January 17, 1893, that the Hawaiian kingdom was "hereby abrogated," and on February 1 an eager Stevens declared
Hawaiʻi to be a U.S. protectorate because, as he informed his superiors in the nation's
capital, "the Hawaiian pear is now fully ripe and this is the golden hour for the United
States to pluck it." But the new president, Grover Cleveland, replaced Harrison before the Senate could act on the treaty of annexation, and on March 9, 1893, he with-
drew it from the Senate and two days later dispatched James H. Blount to investigate
"all the facts you can learn respecting the conditions of affairs in the Hawaiian Islands,
the causes of the revolution by which the Queen's Government was overthrown, the
sentiment of the people toward existing authority, and, in general, all that can fully
enlighten the President touching the subject of your mission." Stevens was relieved
of his post, and after reading the Blount report the secretary of state recommended
against U.S. annexation, admitting that a "great wrong [was] done to a feeble but independent State by an abuse of authority of the United States." Cleveland agreed and
declared that the United States could not annex Hawaiʻi, having achieved that end
through "unjustifiable methods" involving "an act of war . . . without authority of Congress" and the "lawless occupation of Honolulu under false pretexts by the United
States forces." Further, he sought to repair the "substantial wrong" by seeking to restore the kingdom to its former status.[41]

The provisional government had other ideas, and its president, Dole, rebuked the
U.S. president for interfering in "our domestic affairs." "I am instructed to inform you,
Mr. Minister," Dole wrote to Stevens's replacement, "that the Provisional Government
of the Hawaiian Islands respectfully and unhesitatingly declines to entertain the
proposition of the President of the United States that it should surrender its authority to the ex-Queen."[42] After holding a constitutional convention, boycotted by an
overwhelming majority of Hawaiians, the provisional government, on July 4, 1894,
decreed a republic, and six months later, after a failed attempt to restore the monarchy,

FIGURE 24. In this "rescue" and "liberation" of another dark race, Miss Columbia assists a "blackened" Hawaiian girl in this racialized and gendered depiction on the cover of *Judge*, March 25, 1893. Artist: Victor. Courtesy of Bishop Museum, Honolulu.

191 Hawaiian patriots were tried before a military tribunal for the "insurrection," most receiving prison sentences of one to five years and more than two dozen exiled to the United States and Canada. The queen also faced arrest and was convicted of conspiracy and sentenced to pay a $5,000 fine, spend five years in prison, and perform hard labor. Instead, she was held under house arrest, and her movement was restricted for nearly two years.[43]

At her trial, testified Lili'uokalani, the intent was "to humiliate me, to make me break down in the presence of the staring crowd." The judge advocate referred to her as "the prisoner" and "that woman" with "such affectation of contempt and disgust," the queen recounted, and throughout she refused to respond to "their taunts and innuendoes, and showed no emotion." Her sentence, she correctly believed, was not intended to punish her but was staged to teach a lesson of power and control, "to terrorize the native people and to humiliate me." After having served her time under confinement, with the government's permission, the queen left the Islands for the United States to "forget [my] sorrow" and petition President Cleveland for the restoration of her kingdom on behalf of the "patriotic leagues of the native Hawaiian people," unlike the aliens who attempted "to defraud an aboriginal people of their birthrights."[44] Those impostors, the queen pointed out, claimed to be "Americans" when calling upon the United States to protect their interests, as when they conspired with Stevens to deploy U.S. marines, and assumed the name "Hawaiian" to establish their credentials in Washington as legitimate representatives of the people and nation. "They are not and never were Hawaiians," Lili'uokalani rightly declared.[45]

The queen paid visits to Boston and New York City, places where her people had labored and died decades before her sojourn on the East Coast that winter of 1896 and 1897. Those included Hawaiian sailors who manned New England traders and whalers and some who fought in the War of 1812, Hawaiian students who attended a mission school with other native peoples of America and a few Chinese, Hawaiians who enlisted in the army and navy and fought for freedom in African American regiments during the Civil War, and Hawaiians who settled in Massachusetts, Connecticut, and New York.[46] The queen also spent time in Washington, where she socialized with the president and members of Congress and completed her version of the *Kumulipo,* the Hawaiian creation epic and genealogy that tied her and her people to the land as the

FIGURE 25. Festooned in feathers like America's native peoples and barefooted, indicating a "savage," a black-face Lili'uokalani appears supported by U.S. bayonets, though in fact those arms assisted in ending her rule and kingdom's sovereignty. From the cover of *Judge,* December 2, 1893. Artist: Victor. Courtesy of Bishop Museum, Honolulu.

FIGURE 26. Queen Liliʻuokalani, 1913. The queen died at the age of seventy-nine on November 11, 1917. In her diary on the day she learned of the vote in Congress approving Hawaiʻi's annexation, she wrote: "Aue! My love for my birthland and my beloved people. Bones of my bones, blood of my blood. Aloha! Aloha! Aloha!" Photograph courtesy of Hawaiʻi State Archives. Quote from Noenoe K. Silva, *Aloha Betrayed* (Durham, N.C.: Duke University Press, 2004), 199–200.

original children of the sea, land, and skies of Hawai'i. There, she also published an edition of her widely known and loved composition and song, *Aloha 'Oe* (Farewell to Thee), to give to friends and supporters.[47] It was a fitting place, the U.S. capital, for that poetic finish.

Lili'uokalani's sojourn took place while Congress considered the resolution of annexation submitted by President William McKinley, who had succeeded Cleveland in 1897. In protest, the queen listed for the State Department her indictment: "I declare such treaty to be an act of wrong toward the native and part-native people of Hawaii, an invasion of the rights of the ruling chiefs, in violation of international rights both toward my people and toward friendly nations with whom they have made treaties, the perpetuation of the fraud whereby the constitutional government was overthrown, and, finally, an act of gross injustice to me."[48]

But the deed was done, with a president and Congress bent on tropical empire, having gained Cuba, Puerto Rico, Guam, and the Philippines as the spoils of war with Spain. Anna Dole, wife of the republic's president, described Lili'uokalani's return to the Islands on August 2, 1898, after Hawai'i's annexation by the United States about a month earlier: "The ex-queen arrived from the United States where she caused much trouble for the government," she confided to her sister; "but we have won," she declared triumphantly. "Sanford and I went down to the wharf to see what reception she would have. It was strange because it was so quiet. Nobody cheered." But when the queen "raised her hand and said Aloha! A great shout of Aloha came from the crowd. But that was all. Most people were crying."[49]

MISSION BOYS

In 1893, Lorrin A. Thurston, one of the chief conspirators against the kingdom, enumerated the advantages of a U.S. annexation of the Sandwich Islands in the pages of the *North American Review*. Along with the ties of tradition and blood with Hawai'i's white settlers, Thurston reminded his readers, and the strategic position of the archipelago and its mid-Pacific harbors, the chains of commerce bound the temperate, industrial Northeast with those distant tropical isles, profiting both. The 1876 Reciprocity Treaty, he claimed, resulted in "direct financial advantages accruing to the United States and its citizens," which "have more than repaid, dollar for dollar, all loss by the United States through remission of duties." Since the treaty, Thurston detailed, imports from the United States had grown to more than $47 million, American settlers owned $23 million in sugar property, and exports to the United States had reached more than $5 million. Investments in sugar equipment, roads and rail, and shipping had increased, and most of the gains fell into American hands. In fact, he declared, U.S. annexation was a matter "involving the fortunes of thousands of their own flesh and blood, and millions of dollars worth of American property." "The question of what the future policy of the United States towards Hawaii shall be is no longer one in which political advantage to the United States and financial advantage to Hawaii are the only factors, as was the case in 1876; for a trade has been built up," he recalled, "property acquired, and interests have become vested, which make the financial effect of the future American policy of more importance to Americans than it is to Hawaiians." That astonishing claim was made by a leader of the "Hawaiian" League and by one who was "Hawaiian by birth."[50]

From a global as opposed to local perspective, however, Thurston had stated a truth. Hawai'i's tropical plantations, designed by and for Americans, were of greater moment to the United States than to its possession. In addition to the immense fortunes realized, the Islands anchored U.S. claims to the Pacific. Like a flotilla of armed vessels, the archipelago pointed the way to and posed a line of defense against Asia. And insofar as Europe's manifest destiny turned on its representations of and material accumulations from Asia, the Atlantic and Atlantic world were mere preludes to the Pacific and Pacific world. Although not joined by a Northwest Passage, the oceans formed correspondences that the continental divide could not keep apart. America's metals and furs, extracted by its "Indians," enabled access to Asia's wealth, and the Atlantic plantations found fertile ground in the Pacific, cultivated at first by the "Indians" of both oceans. The patterns of land tenure and labor migration in the Atlantic and Pacific bore the impress of tropical plantations, begun in the Mediterranean, as well as the decimation of native peoples, the usurpations of settlers who turned "native," and the impoverishing of airs, waters, and sites to enrich alien peoples, lands, and cultures.

A New Englander like United Fruit Company's Andrew Preston, James D. Dole arrived in Hawai'i in November 1901 after the kingdom's annexation by the United States. His father's cousin, Sanford B. Dole, had served as president of the short-lived republic in 1894, and with the lowering of its flag he became governor of that U.S. territory in 1900. Like Sanford, James was descended from a line of Congregational ministers, and because "the mission boys," in Thurston's words, stuck together for the common good, the newcomer was assured entry into the circle of the anointed, despite landing with only $1,500 in his pocket.[51]

Moreover, in advance of his departure for Hawai'i, James's father, Sanford's close confidant, wrote to him asking about employment for his son. In Boston, James met Walter F. Dillingham, heir to lands and industries in Hawai'i, and a family friend promised to introduce him to the Castles of Castle & Cooke for a job on one of that factor's sugar plantations, as was first suggested by the governor. James, a Harvard graduate, had every reason to anticipate that life in the Islands would be "just one long sweet song," as Dole would recall years later. "I had the idea that after two or three years of reasonable effort expended on cheap government land, I would be able to spend the rest of my life in a hammock, smoking cigars rolled from tobacco grown on my own place, and generally enjoying a languorous life of ease and plenty."[52]

He thought about investing in coffee at first but was persuaded instead to turn to pineapples. "I think I got quite a little excited," Dole recalled of that epiphany and moment, "when I began to vision plantations and canneries and native workers and ships carrying cargoes of fruit to all the world. Perhaps the romantic nature of the pineapple hit me, don't you think? You know it has a personality; an 'IT.'"[53] And thus, beneath a silvery moon and with tropical breezes pregnant with the scent of flowers, James D. Dole began his romance of the pineapple.

7

Hawaiian Pine

As Honolulu's Chinatown lay under medical quarantine, James Dole "spent much of his time loafing on the beach," according to a biography as told by his grandson. And Dole watched as the fire department started the fire that was intended to cleanse Chinatown of the bubonic contagion but that burned out of control, leaving thousands homeless. From the comfort of cousin Sanford Dole's Honolulu home, James celebrated "Inauguration Day, Territory of Hawaii" on June 14, 1900, commencing the U.S. territory and Dole's term as governor. "June 14 was a gala day in the Islands," James wrote in his diary. "In the morning the Inauguration exercises took place. . . . Cousin Sanford in a Prince Albert and stovepipe looked finely and read his address well. . . . In the evening, I stayed out and appeared at the Inauguration Ball, which was quite a swell affair."[1]

A houseguest of Anna and Sanford Dole for several months, James sold articles he wrote on Hawai'i to magazines on the continent, invested in Island sugar plantation stocks, and upon the advice of his well-heeled cousin visited sugar plantations and met with directors of Big Five companies in search of a job. James, however, preferred self-employment over working for others, and he received a tip from the governor, who as president of the republic, had signed the Land Act of 1895, which set aside some 1,300 acres of government land at Wahiawā, O'ahu, for settlers. A 61-acre tract of that homestead land was to be sold in 1900, Sanford was told by his former commissioner of agriculture under the republic, and the governor alerted James to that prospect. In anticipation of the land sale, James appealed to his father for a $2,000 loan, which he received, and he secured the property for $4,000. His neighbors were mainly Californians who formed an agricultural cooperative, the Hawaiian Fruit and

Plant Company, to produce fresh fruit and vegetables for markets in the Islands and on the West Coast.[2]

DIRT FARMER

Hawaiʻi's Chinese, like those in the American West, excelled in raising garden vegetables. In the Islands, they dominated export crops such as rice and were important growers of coffee and bananas, and they supplied most of the fruits and vegetables for the local market throughout the late nineteenth and early twentieth centuries.[3] Dole, joined by Harvard classmate George Damon Dutton, decided not to compete with Chinese farmers by producing instead "poultry and pineapples, peas and other vegetables which the Chinamen do not raise successfully." So in 1900 the gentlemen farmers planted watermelons, potatoes, grapevines, and fruit trees, including apples, avocados, bananas, limes, lemons, oranges, peaches, pears, and star fruit. After discovering that the soil favored pineapples, they concentrated on growing that fruit and began by putting to root several acres. Back in Boston, Dole's parents enjoyed the "wonderful pineapple pickles" made from the first harvest.[4]

The need to preserve pineapples for distant consumers suggested to Dole the idea of canning the fruit. As he recalled of that thought: "So I reached the conclusion that establishment of a cannery at Wahiawa was of prime importance, if pineapple culture was ever to amount to anything, and with this idea came the really important one that once pineapple were in cans there was the whole world for a market."[5]

The U.S. canning industry grew during the nineteenth century on the East Coast, and its primary stimulus was food spoilage and the time required for transportation from the site of production to the marketplace. Boosts to the industry included far-flung wars and the necessity of supplying troops in the field, westward expansion and the California gold rush before agricultural development in the West, and the importation of tropical fruit.

In Hawaiʻi, pineapple canning probably began in Kona, on the island of Hawaiʻi, in 1882, and although the *Honolulu Commercial Advertiser* found the pack to have "excellent flavor" with qualities that would "take first place in any market," the experiment failed as a commercial venture.[6] More successful was John Emmeluth from Cincinnati, who began canning pineapples in 1889 on Oʻahu and shipping them to Victoria, British Columbia, San Francisco, New York, and Boston. Three years later, he joined horticulturalist John Kidwell to form the Hawaiian Fruit and Packing Company, which maintained a pineapple plantation at ʻEwa and a cannery near Waipahu. With its fields planted with several hundred thousand shoots, the company shipped five thousand cases from the first harvest, and by the time it was sold in 1898 the company had marketed fourteen thousand cases of pineapples and four hundred thousand fresh fruit.[7]

By contrast, James Dole's Hawaiian Pineapple Company, capitalized in 1901 with $16,240, was small, but his timing and connections were impeccable. Albert Francis Judd Jr., grandson of missionaries, Yale graduate, and son of a chief justice of the kingdom and territory, agreed to draw up the articles of incorporation and serve the company as its first president. Investors in Hawaiʻi, Boston, and California underwrote the company, which expanded its operations from 10 acres of pineapple leased from Dole to an option to lease 300 acres from a company managed by Walter F. Dillingham, whom Dole had met in Boston.[8] Another contact, cousin Sanford Dole's domestic servant Ah Lim, in apparent gratitude for a loan from James, supplied Chinese laborers for his Wahiawā pineapple fields for wages of about $22 a month.[9] The company's cannery packed 70 tons, or 1,893 cases, of pineapple its first year, 1903, and 8,818 cases the next year. That total leaped to 25,022 cases in 1905, increased to 31,934 cases in 1906, 108,600 cases in 1907, and a remarkable 225,320 cases, more than half the entire output of the Islands' canned production, in 1908.[10] James Dole was no "dirt farmer," as he had claimed,[11] and was rapidly becoming an industrial magnate.

Dole's bet, that Hawaiian canned pineapple could turn a profit, despite previous failures, was helped by U.S. annexation and the consequent dropping of duties on Island products. Without the tariff, Hawaiian pineapples gained advantage over those grown in the Bahamas and West Indies though canned in Baltimore. Further, land and labor were cheaper in the territory than in Florida, another of Hawaiʻi's competitors, and Hawaiian pineapples, its boosters claimed, were superior in taste and texture to those produced in other parts of the world. Finally, Dole expanded his distribution in the United States by signing as his sales agent California's Hunt Brothers, "one of the most successful canners of California, and proprietor of what is probably the second largest fruit cannery in the world," in Dole's words.[12] With friends in influential places, reduced costs, a superior product, and access to potentially every grocery store and home in America, James Dole was well positioned from his tropical anchorage to infiltrate the "inner dikes" of the temperate zone.

In 1903, Dole was elected president of his Hawaiian Pineapple Company, and he set out to build "the largest and most complete pineapple cannery in the world, as soon as production justifies it." That opportunity arose in 1906 when, with increased consumer demand for the fruit, the American Can Company built a factory in Honolulu. To get closer to that supply and the shipping harbor, Dole closed his Wahiawā cannery and built a new facility with warehouses next to the can manufacturer. Dole's company employees invented machines to streamline the pineapple's canning, including a feeder and slicer to cut the fruit into uniform shapes and sizes, an "eradicator" to scrape the skin clean of its flesh and juice, and, most important of all, the ginaca, named for its inventor, Henry Gabriel Ginaca, which removed the pineapple's rough skin, cored and sized its body, and sliced the flesh. The ginaca was capable of delivering thirty-five pineapples per minute to an assembly line of trimmers, usually women who sat at tables to inspect and remove any imperfections on the fruit before sending it for canning. With improvements, the ginaca became the industry's standard, capable of handling more than one hundred pineapples each minute.[13]

Efficiency, which lowered expenses and increased production, was thus enhanced by investments in research and development, Dole having secured by 1921 twenty-nine patents on pineapple production. The pineapple's ends and shell, about half the processed fruit, were at first discarded—until researchers found that the waste could be shredded, pressed, and dried to produce bran to feed pigs, cattle, and horses. By 1927, more than 7,000 tons of pineapple bran was sold to livestock producers. Dole's cannery operated day and night during the harvest season, and it set a world record for fruit canning in 1926, when it produced 76,693 cases in a single day. In 1930, with an expanded cannery of 1.6 million square feet, nearly 37 acres of floor space, it set a new record of 94,085 cases.[14]

A fully integrated agricultural industry, like sugar plantations, Dole's pineapple production included the land on which the fruit was grown, a pool of cheap but

FIGURE 27. In the modern industrial assembly line, humans tend machines that ensure uniform products. The ginaca, invented in 1913, revolutionized pineapple processing. Courtesy of Hawai'i State Archives.

efficient labor, a cannery for processing and packing, a department for research and development, and agencies for advertising, sales, and distribution. Starting with the insider's information that led to the acquisition of his Wahiawā homestead plot, Dole's real estate dealings were charmed. He negotiated additional lands from an influential contact, Walter F. Dillingham; his wife's uncle, a pineapple grower on Maui, may have had a role in Dole's purchase of a 51 percent interest in the Haiku Fruit Company on that island in 1909. Together, the two companies held a 60 percent share of the territory's total pineapple output. In 1922, in control of 9,400 acres of pineapple fields that yielded 48,000 tons of pack, Dole arranged a deal with one of his principal landlords, the Waialua Agricultural Company, which was partially owned by

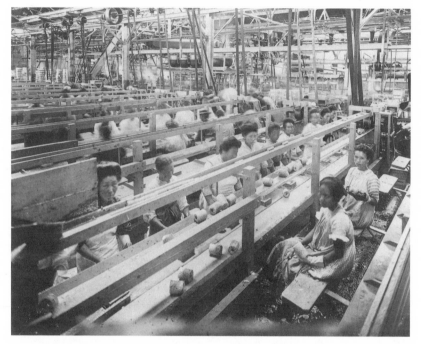

FIGURE 28. Women, under careful supervision, rectify blemishes on processed pineapples missed by machines. Courtesy of Hawai'i State Archives.

Castle & Cooke, for 12,000 acres and $1.25 million in exchange for a one-third interest in Dole's Hawaiian Pineapple Company. With the cash, Dole bought virtually the entire island of Lana'i from Maui ranchers the Baldwins, missionary descendants, and thereby possessed nearly 45 percent of the best pineapple lands in Hawai'i. In 1930, with 25,143 acres of pineapple under cultivation, Dole's company packed nearly 4.5 million cases of the fruit, representing a tonnage increase of more than 40 percent.[15]

"We bought Lanai to get room to grow in," remarked Dole of his purchase, which followed a search in Mexico, the Dominican Republic, Fiji, the Philippines, and Australia for new pineapple lands. Having visited and rejected those places, Dole pro-

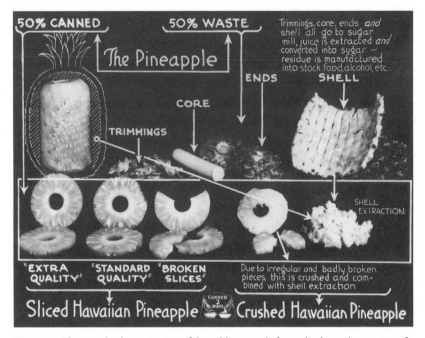

Figure 29. Charting the domestication of the wild pineapple for civil palates, the creation of a rank order of "quality," and the utilization of all of its parts. Courtesy of Hawai'i State Archives.

ceeded to transform Lana'i into the "Pineapple Isle."[16] Friend and patron Walter F. Dillingham's Hawaiian Dredging Company built the harbor; the island's Hawaiians were displaced by Filipino, Chinese, and Japanese laborers imported mainly from O'ahu and Hawai'i; and when Lana'i's "open house" was held on January 30, 1926, Dole chartered a steamship to carry to his island his guests, including the territorial governor, Honolulu's mayor, the commanding general of the Hawaiian Department, the University of Hawaii president, bankers, newspaper publishers, and others.[17] The "Isle of Pines," the February 1, 1926, *Honolulu Advertiser* declared on its front page, was an "epic in development." Hawai'i's "smallest Isle" was "home of the greatest pineapple activity in the world." Against that "development," Hawaiians on Lana'i declined from about 2,500 in 1823 to fewer than a hundred in 1902[18] and "experienced the disheartening realities of losing their native lands and land rights" and "began to feel pressured by foremen and customs sometimes contrary to their traditional behaviour and beliefs."[19]

DISCIPLINING WORKERS

With shared business interests beyond canned pineapple's need for sugar in its syrup, the pineapple growers drew at first from the sugar planters for their excess labor of mainly Asian or "foreign races," in the words of a Labor Department report, to tend their fields and factories.[20] By 1915, pineapple was second only to "king sugar" in Hawai'i's economy. Although there were some bases for cooperation, the conditions and patterns of labor differed in those two industries. Pineapple cultivation was less labor intensive than that of sugar, and its production was largely seasonal, concentrated in the summer months from June through August. Its labor force, thus, consisted of a small group of year-round workers, about one-sixth the total during peak times, and a large group of intermittent and seasonal laborers for the harvest and cannery pack. The latter consisted of women and schoolchildren, who received lower wages and fewer benefits than the men who formed the core of permanent workers.[21]

Race and gender affected the daily wage, as shown in table 1. In the field, Asian, Hawaiian, and Portuguese children earned an average daily wage of sixty-three cents; in the cannery, Asian, Hawaiian, Portuguese, and Puerto Rican children earned an average of fifty-two cents.

Further, before 1915 the industry relied heavily upon independent producers, especially Japanese, whose small farms supplied fruit for the canneries.[22] Thereafter, companies like Dole's expanded their fields and plantings to control the supply and its costs, phasing out the independent farmers and consolidating their hold over the industry, which came to resemble sugar in its concentration of capital, such that over a thirty-year period beginning in 1903 three of the largest companies controlled 70–85 percent of the Islands' total pineapple pack.[23]

Increasingly, mechanization replaced workers while requiring skilled laborers to maintain and operate the equipment and to undertake certain tasks that machines

TABLE 1. Pineapple Labor Force in Hawai'i, 1915

| Race | FIELD | | CANNERY | |
	Number of Workers	Daily Wage (average)	Number of Workers	Daily Wage (average)
American (m)	3	$2.00	23	$1.26
Chinese (m)	312	1.02	218	1.07
Chinese (f)	—	—	154	.72
Filipino (m)	223	.93	669	1.06
Filipino (f)	2	.55	141	.65
German (m)	—	—	2	2.11
German (f)	—	—	2	.85
Hawaiian (m)	85	1.07	143	1.11
Hawaiian (f)	22	.56	204	.70
Hindu* (m)	2	.90	—	—
Irish (m)	—	—	1	2.00
Japanese (m)	632	1.02	897	1.10
Japanese (f)	132	.64	557	.69
Korean (m)	123	.95	297	1.11
Korean (f)	1	.60	106	.68
Mexican (m)	—	—	1	1.19
Part-Hawaiian (m)	8	1.26	12	1.32
Part-Hawaiian (f)	—	—	87	.76
Porto Rican* (m)	14	.97	7	.94
Porto Rican* (f)	—	—	7	.69
Portuguese (m)	82	.99	62	1.06
Portuguese (f)	15	.63	103	.72
Russian (m)	1	.90	—	—
Russian (f)	—	—	35	.70
Spanish (m)	15	1.01	16	1.06
Spanish (f)	2	.68	54	.68
Race unknown (m)	—	—	152	1.03
Race unknown (f)	—	—	81	.73
Total and average, male	1,500	$1.08	2,500	$1.24
Total and average, female	174	.61	1,531	.71

SOURCE: U.S. Bureau of Labor Statistics, *Fifth Annual Report of the Commissioner of Labor Statistics on Labor Conditions in the Territory of Hawaii, 1915,* Sen. Doc. 432, 64th Cong., 1st sess. (Washington, D.C.: Government Printing Office, 1916), 45.

*The spelling "Porto Rican" and the attribution "Hindu" are in the original report.

FIGURE 30. Backbreaking stoop. Skilled laborers plant suckers, slips, and crowns on rows of paper mulch. Courtesy of Hawai'i State Archives.

were incapable of performing. So, although the preparation of the field for planting was completed by machines, planting and harvesting still required work crews. And because pineapples ripen at variable rates, men and women selected and picked the fruit by hand and hauled them in sacks to animal-drawn carts and later trucks on roads bordering the fields.[24]

A boom with a conveyor belt that extended over several rows of plants eventually replaced this labor-intensive harvesting method. Men in rows followed the boom and, spotting a ripe fruit, removed it from its stalk, twisted off the crown, and flicked the body onto the belt, which carried the pineapples to waiting trucks. Outfitted with lights, the boom enabled harvest day and night during the season's peak.[25] A *luna,* or overseer, often stood on the harvester to monitor the pickers to ensure against their missing or bruising the golden fruit, and he set the cadence of the machine's march through the field.

FIGURE 31. A watchful *luna* on horseback and pickers. Courtesy of Hawaiʻi State Archives.

The men and women in the fields and canneries were associates of machinery, which may have liberated them from backbreaking labor but also served to discipline them. The machine's rhythms and abilities calibrated the speed and tasks of humans, who were bound to them in a relentless drive for production. As told by Ida Kanekoa Milles, a trimmer at Dole's Hawaiian Pine in 1946: "Well, when they hired me as a trimmer, I didn't have any idea that it's going to be hard, or I'm going to be wet with juice on my apron. To tell you the truth, I sat on that table and looking at the pineapple coming through the Ginaca [machine] to me look like 100 pineapples a minute. Before I finish one pine, there's another one coming down. I put this pine down, I pick up another. I didn't finish, another pine coming down. Before I realize, there was a pile of fruit in front of me."[26]

That singleness of purpose, of machines and humans, mirrored Dole's desire to have his workers identify with the company. "I have been particularly interested in

FIGURE 32. Heavy protective gear and clothing, heat from the sun and engines, and dust from the red dirt and plant leaves conspire to sever the union between pickers and the harvest boom in the machine's relentless march through the field. Courtesy of Hawai'i State Archives.

trying to organize our business in such a way that every employee, so far as possible, may feel that his interest is that of the company and vice versa," he wrote. And like the sugar planters during the interwar period of paternalism, Dole planned model communities in which his workers lived, such as Lana'i City, and dispensaries for the sick and injured, cafeterias and day-care centers, and sports facilities. "The payment of good wages and providing safe, healthful and morally wholesome conditions for the work in the factory and on the plantations" were Dole's primary goals, according to his biographer. As a result, boasted a report by Castle & Cooke, "Hawaiian Pine was the first successful pineapple canning enterprise in the world . . . (and) the leader of the development of the industry."[27]

At the same time, a Labor report noted, "even in the best of the canneries there was a tendency here and there to crowd in extra girls at the [packing] tables, so that in spite of a prevailing sense of space and roominess, the girls were packed in quite closely and barely had elbow room," and "endless belts carry the pineapple cylinders past rows of women inspectors—white-capped, aproned, and rubber-gloved—who examine the fruit and cut out with sharp knives all particles of shell or foreign matter that the ginaca did not reach."[28] In sum, a Labor Department official observed in 1939, "the canning industry is highly mechanized and there is little heavy work. Most of the women's jobs are simple, and dexterity and speed rather than skill seem to be the prime requisites."[29] Like the *luna* in the field, the "forelady" in the cannery monitored the assembly line. As told by Kanekoa Milles, who gained promotion: "The pay was better and the cap was different, too. Forelady is a blue cap. I'm responsible for fourteen to twenty-four girls. I have to watch that the girls are doing the proper way: handling, trimming, and putting their fruit on the chain." Although clean and well lit, the cannery, operated primarily by women and "workers whose racial descent is other than Caucasian,"[30] was designed by its owners to be an instrument of mass production, uniformity, and control.

Paternalism, on sugar and pineapple plantations, cut at least two ways—as a work of charity and as a regime of regulation. Each served its master by promoting production and retarding discontent and strife. One of Dole's managers on Lana'i during the 1930s, Harold Blomfield-Brown, a social historian related, "was perhaps the most colorful manager of all." Blomfield-Brown allowed no domestic animals or pets on the island, and from his house on top of a hill he watched "every activity on the plantation from a telescope." When spying workers loafing on the job, he "rode into the fields to give them a tongue-lashing. If he caught anyone dropping candy wrappings or other trash in the village, he personally inflicted a strong reprimand. When the inter-island boat docked at the harbor at Kaumalapau, [Blomfield-] Brown, impeccably dressed, would inspect each debarkee to keep out gamblers, salesmen, and prostitutes."[31] A 1930 article described the Lana'i manager, the "general ruler" of "his kingdom," as "genial, hospitable, skilled in the handling of orientals."[32] Workers, as dependents, required the supervision of *lunas* in the fields and supervisors in the canneries, and as self-contained communities, plantations made so little distinction between work site and home that managerial regulation reached into the private sphere.[33]

Workers, however, were not easily regimented, and most sought to improve their conditions despite seasonal jobs, which tended to reduce the stakes for those workers, making them more difficult to organize and less likely to contest workplace inequities. But as early as 1914, cannery workers struck for higher wages, and three years later another group of cannery laborers went on a five-day strike over wages.[34] Even employers, including Dole, recognized and tried to take advantage of worker agencies by recruiting striking workers during labor disputes such as the historic strikes on O'ahu's sugar plantations in 1909 and 1920, when Dole placed full-page advertisements in Japanese-language newspapers offering employment in pineapple to sugar strikers.[35]

In 1951, on Dole's Lana'i plantation, by then owned by Castle & Cooke, workers struck over wages and general grievances, including the poor condition of camp housing, the arrogance of company officials, and the lack of dignity for workers and their families. "Not a single bit of that island is owned by the employees," Pedro de la Cruz, the strike leader, observed of Dole's creation, the "Pineapple Isle." "Every damned thing, our homes and everything belongs to the employer. Only our lives belong to us." For 201 days, the 752 laborers stayed away from fields rotting with pineapples. Facing the prospect of devastating losses, the company finally capitulated to the union's demand for an industrywide wage increase.[36]

ADVERTISING MARKETS

Besides securing land and labor and canning its fruit, the fully integrated pineapple industry promoted sales while carving out a market where virtually none had existed before. Dole was a leader in those aspects of the business. After conferring with agent Hunt in San Francisco, Dole proposed the Hawaiian Pineapple Growers Association, which, from 1908 to 1912, promoted Hawaiian pineapple on the continent. Conceived of as a way to unite independent and competing pineapple producers for their collective gain, the "cooperative association of advertising" sought to market pineapples as a "generic" food item by associating it with an image of Hawai'i and not individual brands. The joint advertising campaign, accordingly, was to "make the word 'Hawaiian' mean to pineapple what Havana meant to tobacco."[37] In that way, by selling "Hawaiian pineapple," all of the association's members would benefit by capitalizing upon Hawai'i's carefully crafted image of a lush, tropical paradise.

Publicists for the Hawaiian Pineapple Growers Association, based in New York City, had to work with and against commonly held views. Island paradises beckoned, but prejudices against canned foods had to be overcome, along with a widely held belief that pineapples were "so tough and so stringy; it bites the tongue and actually makes the mouth sore," as is admitted in an advertisement in the January 1909 issue of the *Ladies' Home Journal.* Accordingly, the text emphasizes that "Hawaiian Pineapple Is So Different." The "Hawaiian" is "the best variety of pineapple this earth ever produced," the sales pitch purrs, "the flesh is tender without a trace of woody fibre; the flavor rich, yet delicate, and without a suggestion of the disagreeable 'bite' which makes all the fresh pineapple that comes to our market so disappointing." Picked only when fully ripe at the peak of perfection, the "Hawaiian Pineapple contains nothing but fresh fruit and pure granulated sugar. It is put up only in sanitary cans preventing contamination by solder or acid. No human hand touches the fruit in peeling or packing." And just in case the homemaker was at a loss over how to serve the canned pineapple, the advertisers offered *Hawaiian Pineapple,* a booklet containing "tested recipes for this most excellent of all preserved fruits."[38]

Successors of the Hawaiian Pineapple Growers, the Hawaiian Pineapple Packers' Association and the Association of Hawaiian Pineapple Canners, continued the advertisement campaign that sold a generic product, Hawaiian pineapples, and expanded its market by offering easy, free recipes for busy housewives. As an advertisement in the March 1925 *Ladies' Home Journal* noted, sliced pineapples were perfect for "serving right from the can and for quick desserts and salads," and the versatile crushed variety could be used to create sundaes, ices, pies, cake filling, salads, and "hundreds of made-up dishes." Moreover, those "golden circlets of tropical goodness," the text tantalized, would make you forget winter and transport you to "a vision of balmy ocean breezes, rustling palms and a lazy surf, washing coral shores."

In the September 1925 *Ladies' Home Journal,* the Association of Hawaiian Pineapple Canners, based in San Francisco, offered $50 for "the very best recipes that

FIGURE 33. The sanitary can and immense fruit, foregrounding palm trees and calm waters, promising modernity and a taste "so different." *Ladies' Home Journal,* March 1, 1909, 66.

Hawaiian Pineapple

**PickedRipe—
Canned
Right**

The full flavor,
quality and
tenderness of
ripeness, from
the fields of
Hawaii to
your table.

Cuts
with a
spoon like
a peach

Drop postal
for this

Free
Book

Better
than home
preserved pineapple
or the best pineapple on the market.
It's so different, because the pineapples you
buy are picked green and ripened afterwards

on "How to Serve Pineapples."
Handsomely illustrated — many tested recipes

Hawaiian Pineapple Growers' Association, Tribune Bldg., New York City

FIGURE 34. In its third month of advertising, the long text is reduced to its essence, "Picked Ripe—Canned Right." *Ladies' Home Journal,* April 1909, 72.

the women of America have discovered" for Hawaiʻi's "king of fruits" and "one of America's most popular fruits." And in the April 1926 *Journal* issue, the association announced the one hundred winners from sixty thousand recipe submissions and published them in a free booklet, *Hawaiian Pineapple as 100 Good Cooks Serve It.* Winners included Mrs. Thelma Cox of Pittsburgh, Pennsylvania, whose "ice-cream pie—Hawaiian" promised to appeal to "every woman who enjoys serving novel dishes"; Mrs. W. T. Rowland of Charlotte, North Carolina, whose "water lily salad" won in the salad category; and Miss Mary Deschamps of Brooklyn, New York, whose pineapple and marshmallow dessert graced "more than one company dinner."[39] Each of the women, the booklet noted, "is a practical home cook. Their recipes—they assure us—were developed to meet the needs of average home service. Try them! Let them add variety and sparkle to *your* menus! You will find that they open up a host of pleasing possibilities for serving this 'King of tropical fruits.' "[40]

The campaign pioneered more than

FIGURE 35. By June, the stress is on the canned pineapple's versatility, in simple to elaborate dishes. *Ladies' Home Journal,* June 1909, 63.

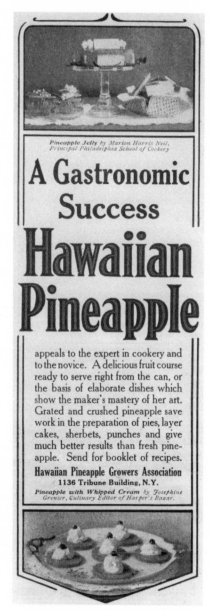

Pineapple Jelly *by Marion Harris Neil, Principal Philadelphia School of Cookery*

A Gastronomic Success
Hawaiian Pineapple

appeals to the expert in cookery and to the novice. A delicious fruit course ready to serve right from the can, or the basis of elaborate dishes which show the maker's mastery of her art. Grated and crushed pineapple save work in the preparation of pies, layer cakes, sherbets, punches and give much better results than fresh pineapple. Send for booklet of recipes.

Hawaiian Pineapple Growers Association
1136 Tribune Building, N.Y.

Pineapple with Whipped Cream by Josephine Grenier, Culinary Editor of Harper's Bazar.

generic advertising; it was among the first to tie recipes with product sales to extend its market reach.[41] Ordinary homemakers required instruction on the use of this tropical fruit, both fresh and canned, whose familiarity was mainly confined to select classes. For instance, a 1931 issue of the *Ladies' Home Journal* carried an article that declared, "May is Pineapple Month," and explained how to select ripe, fresh fruit, how to remove the pineapple's rough skin and slice its flesh, and how to serve it as a cocktail and in salads.[42]

With periodic advertising campaigns, Hawai'i's pineapple industry prospered as a whole, but three of the leading producers—Dole's Hawaiian Pineapple Company; Libby, McNeill & Libby, based in Chicago and in Hawai'i in 1909; and Del Monte, or California Packing Corporation, which began selling Hawaiian pineapples in 1917—tried to capture larger shares of the market as distinctive brand names. As early as 1910, "Drink Dole's Pure Hawaiian Pineapple Juice" appeared in the *Ladies' Home Journal,* carrying James D. Dole's signature and the advice, "Be sure this name in red is on the label."[43] Libby advertised "California" asparagus and fruits and "Hawaiian" pineapple in 1916, and Del Monte featured "Hawaiian" crushed pineapple in 1926.[44] By 1931, with brand-name advertising in full swing, Libby claimed "the finest pineapple ever grown" on "Libby's vast Hawaiian plantations," and Dole, describing itself as the "world's largest growers and canners of Hawaiian pineapple," insisted that discriminating consumers "Look for DOLE stamped in the top" of each can.[45]

In the midst of that advertising campaign, Dole's Hawaiian Pineapple Company approached N. W. Ayer & Son, one of the nation's leading advertising firms, to invite renowned modernist artist Georgia O'Keeffe to Hawai'i as its guest for two paintings on any subject but for use in its pineapple commercials. Although O'Keeffe confided to a friend, "At this moment I don't much care" about the invitation, after examining materials on the Islands the artist became "much interested," and days before her departure a fellow painter reported that she was by then "eager for the experience."[46] As early as the late nineteenth century in England and Europe, fine art's association with "high" culture served to elevate the class stature of the products it advertised, from soap to insurance, and twentieth-century corporations in the United States saw patronage of modern art in particular as good for business and public relations.[47] Further, modern art, with its bold designs and colors, appealed to the senses and mind in an act of persuasion. Hawaiian Pineapple perhaps thought it appropriate that the "king of fruits" and a tropical trophy might require recuperation, having been reduced to a mass-produced, tinned food item widely available on grocery store shelves and in the pantries and recipes of ordinary housewives.

Whatever its purpose, the company paid for O'Keeffe's transcontinental and then oceanic passage from San Francisco on the *Lurline,* which berthed in Honolulu Harbor on February 8, 1939. On O'ahu, O'Keeffe visited the pineapple fields, which she described as "all sharp and silvery stretching for miles off to the beautiful irregular mountains. . . . I was astonished—it was so beautiful." When she asked to live with workers in their camp near the fields so she could paint pineapples, the company refused, put her up in a hotel in town, and gave her a "manhandled," in her words, pineapple with which she was "disgusted." She flew to Maui, where, she wrote, "I enjoy this drifting off into space on an island— . . . I like being here and [I'm having] a very good time."[48] From Maui, O'Keeffe went to Hawai'i and Kaua'i, and on April 14, 1939, she sailed from Honolulu to San Francisco and crossed the continent to return to her studio in New York City.

Painting in both Hawai'i and New York City "from drawings or memories or things brought home," O'Keeffe had completed renditions of a papaya tree, a heliconia blossom, and, after protest, a pineapple bud by the summer of 1939. Despite an illness, the artist debuted twenty canvases from her three-month visit to Hawai'i in her February 1, 1940, annual exhibition in New York City. "If my painting is what I have to give back to the world for what the world gives to me," she noted in the text to the exhibit, "I may say that these paintings are what I have to give at present for what three months in Hawaii gave to me." In reality, she observed, the forms on canvas seem "infinitesimal compared with the variety of experience," which becomes "a part of one's world. . . . Maybe the new place enlarges one's world a little. Maybe one takes one's own world along and cannot see anything else."[49]

Regardless, whether projections of her world, as in other European inventions of the tropics, or expansions of her world, two of her portraits, *Pineapple Bud* (1940) and *Heliconia—Crabs Claw Ginger* (1940), helped to sell Dole pineapple juice in the *Saturday Evening Post, Ladies' Home Journal, Woman's Home Companion,* and *Vogue.* An accompanying text formed playful, subliminal associations of Hawai'i with the pineapple, "hospitable Hawaii," and the tropics with its "abundance" of ("fragrant") flowers and ("golden") sunshine. "Hospitable Hawaii cannot send you its abundance of flowers or its sunshine," it teased. "But it sends you something reminiscent of both— golden, fragrant Dole Pineapple Juice."

BRAND NAME

In 1930, shortly after the stock market crash in the fall of 1929, *Fortune Magazine* profiled Dole's Hawaiian Pineapple Company as "one of agriculture's few prosperous exponents." "Pineapples in Paradise," the featured article was titled, or "how the cactus and the hoodoo disappeared from Lanai" and "why Jim Dole traveled 5,000 miles to go into agriculture." Despite widespread skepticism and doubts, the author noted, Dole and his Hawaiian Pineapple Company succeeded in transforming Lana'i into a thriving pine producer and Lana'i City into "an industrial city in which the ideal of a company town has been truly realized."

The heroic figure of James Dole emerges from the account a self-made man who lifted himself up from trying circumstances ("he found the Paradise of the Pacific something less than heavenly") to capitalism's paradise and ideal, "more than prosperous." Having made pineapple second only to sugar in the Islands by carving out a worldwide market for the fruit and identifying the pineapple with Hawai'i, "Mr. Dole is a greater islander than any sugar tycoon." "For the growth and prosperity of the Hawaiian pineapple were almost entirely the work of one man . . . Mr. James D. Dole, president and general manager of the Hawaiian Pineapple Company." In that sense, "it is certainly fair to say that the pineapple business began in 1899 and in that suburb of Boston, Massachusetts, which is known as Jamaica Plain," Dole's home. The wide gulf, thus, between Dole's New England and his Hawai'i, in the temperate and tropical zones, was bridged and hence subdued by his kind, his "stock that sent the clipper ships to the ports of Japan and of China" and "that sent also missionaries to carry the word of the Lord to distant and heathen regions."[50]

The accolade notwithstanding, James Dole, like many other capitalists of his time, failed to survive the Great Depression, when pineapple sales fell and his company's debts rose despite drastic cuts in salaries and employees. Castle & Cooke assumed control of the Hawaiian Pineapple Company in 1932, although Dole continued as chairman of the board, and under new directors the Big Five firm engineered the company's recovery in 1936, when it finally turned a profit. Management dismissed Dole in 1948, and in 1961, three years after his death, it renamed his company Dole Pineapple Company and, later, Dole Foods Company, keeping his name despite having terminated his services. Such was the power of the brand name with the buying public.

In 1957, Henry A. White, executive vice president of Castle & Cooke and Dole's successor as president of the Hawaiian Pineapple Company, delivered an address to the Newcomen Society in North America in tribute to James Dole for "the great role he has played in the development of Hawaii and in furthering the progress of the American food industry." Dole has earned the title of "Pioneer," White began, for laying the foundation of the Hawaiian Pineapple Company and pineapple industry and for "early recognizing the possibilities of an almost unknown food product when few others shared his vision." From a "young Jim Dole, fresh out of Harvard," to the leader of the national campaign to make the pineapple synonymous with Hawai'i, Dole personified the enlightened leader. "His employees found him a warm, human man who loved to get away from his desk for a visit with employees at the cannery and plantations," White declared, and, quoting Dole, "We have built this company on quality, and quality, and quality." "New products have taken their places on the grocery shelves of the Nation," White concluded, "beside the fine Hawaiian pineapple that Jim Dole established there. More new products will follow. All of them will proudly bear the name of Dole, the name of a great builder and pioneer of industry in the Pacific."[51]

At the onset of World War II, Hawai'i supplied more than 80 percent of the world's fresh and canned pineapple. Although the Islands produced the majority of canned pineapple through 1960, they slipped in overall production as the major corporations, including Dole, transferred their operations from Hawai'i to other countries within the tropical band in Asia, Africa, Central America, and the Caribbean. As early as the 1920s, Del Monte started in the Philippines what became the largest pineapple plantation in the world, and Libby explored possibilities in the Philippines and Southeast Asia, South and Central America, the West Indies, and East Africa. With that global spread, pineapple acreage in Hawai'i dropped from 73,000 in 1959 to 62,400 in 1969 and 33,000 in 1989. When Dole closed its Lanaʻi plantation in 1992, pineapple lands in the Islands had diminished to a mere 16,000 acres.[52]

Yet within the U.S. continent and even on the East Coast, where both fresh and canned pineapple enters primarily from the Caribbean and Central America, pineapple and the islands of Hawaiʻi remain a familiar coupling, along with the idea that Hawaiian pineapples remain the freshest, sweetest, and best in quality. And though the Atlantic and temperate zone's industry may have penetrated the Pacific and its tropics, pineapple's sensory allure of sight, taste, and smell emanating from Paradise haunts the temperate core, promising rebirth and rejuvenation and unbridled bodily pleasures and freedoms.

8

Pineapple Modern

The genius of Hawai'i's pineapple producers illustrated in the spectacular rise of the industry and its influence in the U.S. marketplace of consumption and image making derived at core from its partnership with modernity. Pineapple modern emerged from its production in fields planted with a hybrid creation, the Cayenne, carefully selected for its size, shape, and flavor and tended by research scientists, who monitored the precise rows of plants "all sharp and silvery stretching for miles off" against pests and blights. It created for itself machines to reduce the need for stoop labor; the sanitary, efficient canneries and their lines of uniformed and gloved women who nursed the naturally wild, prickly fruit made tame by peeling and slicing it for civilized tongues; and the advertising campaign aimed at white middle-class women and orchestrated by modern artists, graphic designers, and writers whose productions tapped inchoate desires for fashion and style, sensual abandon, and convenience.

IMPROVING NATURE

"The canning of pineapple has demonstrated," a U.S. Department of Commerce report asserted in 1915, "probably a little more forcibly than has been the case with the canning of any other product, certain unusual values of the discovery of the preservation of articles of food in hermetically sealed containers by means of sterilization by intense heat. In the case of pineapples, a tropical fruit that can not stand shipment for long distances, the flavor and bouquet have been retained in the canned article to a surprising degree, and a further advantage to the consumer lies in the fact that the fruit requires no preparation for the table." Once maligned as unhealthful and taste-less,[1] canned foods now duplicated and even improved upon nature, the report claimed, exceeding its limits of space and time. "Thus," the study boasted, "the canning industry has made it possible for the housekeeper to have on her pantry shelves, in the coldest climates, a tropical fruit in nearly its natural state, ready for serving at a moment's notice, in season and out of season, ready to eat without preparation, and always delicious."[2] Besides bringing the tropics into closer communion with the temperate zone, the canning industry and the food processing industry generally helped to convert local and regional cuisines into national and global food cultures.

The canning industry began when the French government offered a 12,000-franc prize in 1795 to the inventor of a means to preserve food for its soldiers and sailors in the pursuit of empire. A Parisian confectioner, Nicolas Appert, won the contest in 1809, and a year later he published his method of sealing food in glass bottles immersed in boiling water.[3] The London firm Donkin and Hall modified Appert's technique by using cans made of iron sheets dipped in molten tin.[4] In the United States, the Civil War promoted canned goods, which were staple issues for the Union Army, and the industry grew in Baltimore, where several key inventions in the canning process were made.[5]

Canned foods, then, in France, England, and the United States, accompanied the emergence of industrialization and cities distanced from rural farms and fields and far-ranging military expeditions that required supplies, notably food. Canned goods were reliable sources of nutrition even in the remotest corners of the empire, a physician in the British army testified in the late nineteenth century. "Taking my experience in India and the late Nile expedition in which the test of tinned provisions was exceptionally severe, from continued exposure to the powerful direct rays of the sun," he reported, "I have found that tinned provisions, meat, and vegetables, put up separately, or combined in the form of soups, are practically undamageable by any climatic heat."[6] The American Can Company, formed in 1901, the largest manufacturer of cans in the United States and the supplier to Hawai'i's pineapple industry, achieved economic stability only after receiving massive orders from the military during World War I.[7]

Cast as old wives' tales by modernists were notions that canned foods transmitted diseases or that they lacked taste and nutritional value, the science of canning notwithstanding. So while a traditionalist bemoaned "the canned asparagus, pale as death and bitter as gall" and "the sad Brussels sprout, swimming in cool, greasy water," a modernist claimed "that canned asparagus is of a much higher quality than most fresh asparagus; that canned spinach is almost always cleaner than fresh." The traditionalist favored the "strong races," whose "basic fiber" drew from food that was "real, elemental, natural substance" over "the ordinary American, whose tastes are no more discriminating than they could be with a background of canned food and canned education." Among those, the traditionalist complained, "the gospel of machines, speed, and noise dominates the home as well as the factory, office, hotel, and restaurant. So the man humps himself to make a bigger pile, and the woman humps the family stomachs so she may play more bridge." For the modernist, though, canning involved "ingenious machinery," and the industry's rapid growth was an index of development and source of national pride such that "no country competes with us in our consumption of preserved foods."[8]

"The climate and soil of the Hawaiian Islands, together with the success of the Smooth Cayenne variety of pineapple under the conditions prevailing there, has produced a most successful fruit with a flavor as fine as can be found in any section of the world," a 1915 Department of Commerce report noted. "The modern well-ventilated buildings and nearly automatic equipment of the various plants and the small amount of handling during the different processes of canning leave little to be desired in the matter of cleanliness." In fact, the report stressed, "the fruit is practically never touched by human hand from the time it is peeled, the rubber gloves of the sorters being the nearest approach to it."[9]

INARTICULATE LONGINGS

The modern pineapple industry in Hawai'i was a product of its times, the late nineteenth and early twentieth centuries, when modernity and consumer society engrossed and occupied the domestic core and colonialism engulfed the tropical peripheries, a relationship designed to enrich colonial masters and impoverish their colonial subjects. Justifying that exploitative, imperial formation were noble ideas of the social and religious uplift of inferior, premodern peoples and, in the case of the United States and its "new possessions"—the islands of Cuba, Guam, Hawai'i, Puerto Rico, and Samoa—seemingly benign representations of tropical splendor, abundance, and regeneration.

Modernity's culture thrived amidst industrialization, urbanization, a revolution in transportation and communication, and wanton consumerism. The nation eclipsed the village as marketplace with advances in transportation and communication, and emphasis on factories designed for efficiency, including the bonding of humans with machines, led to dramatic increases in production and consumer goods. Capital concentrations abetted the rise of cities, and social and material distances between urban and rural spaces and the North, South, and West marked differential powers, supplies, and needs. Collaborators in those developments were scientific and technological investments and discoveries that enabled and directed the processes of modernization at home and abroad, and they sired grand ambitions of conquests of lands, peoples, and nature itself. It was an age of unbounded optimism and, like the conjured tropics, rampant with fecundity and endless summers.

Insofar as manufacturing was limited by consumption, producers strived to extend their markets by spreading from the local to the national to the global, by targeting white middle-class women as potential consumers, and by installing an all-pervasive culture of consumerism in which appetites were never completely satisfied and where obsolescence required constant renewals. The advertising industry, germinating with and feeding off every new department store, merchandising house, and supermarket, helped to create that modern consumer society.

Fancying themselves as society's trendsetters, advertisers were the "apostles of modernity," as aptly put by one scholar, and "in selling leisure, enjoyment, beauty, good taste, prestige, and popularity along with the mundane product" they assumed that "the customer was 'pre-sold' on these 'satisfactions' as proper rewards for the successful pursuit of the American dream."[10] They, by and large, saw their work as an exercise in persuasion and, insofar as they succeeded in inducing women and men to want the latest in fashion and features for status and convenience or to avoid embarrassment, they affected profoundly the structure and performance of the market and industry, distribution channels, and consumer welfare.[11]

Advertisers commonly gendered themselves as intelligent men and their buying public as ignorant, fickle, irrational, and childish women.[12] An advertising agent wrote that the American woman possesses "inarticulate longings" whereby she "wants it but she doesn't know it—yet!"[13] White middle-class women, "the shoppers of the world" according to an influential 1929 guide for advertisers, were the industry's special targets,[14] mirroring their rise as both producers and consumers, the "new" women of the modern period, who were both housewives and wage workers with buying powers. Further, technology and science entered the home to impose regimes of management and economy, aesthetics and style, health and hygiene, giving rise to a domestic science called "home economics."[15] With housework service oriented and the household a site of consumption, white middle-class woman as homemaker and consumer took center stage in advertising and selling modernity.

Guiding the advertising campaign were two seemingly contradictory attributions of the modern woman, revealing both sides of the gendered coin. On the one hand, "Mrs. Consumer," so named by Christine Frederick, was ruled by her "instincts," including sex love, mother love, love of homemaking, and vanity, or love of adornment. On the other, she operated as a rational decision maker, choosing among products based upon price, quality, and overall value. Advertisers, Frederick advised, should appeal to women's sense, emotions, and feeling, while explaining the benefits of labor- and time-saving devices and goods and instructing her on the sciences of nutrition, wellness, and beauty.[16] And, advertisers repeatedly intoned, modernity required keeping up with the newest findings and products to possess better homes and gardens, more leisure, more pleasure, and greater freedoms from ignorance and drudgery.[17]

DOMESTIC SCIENCE

Commercial and professional interests including food producers and processors, scientists, and advertisers were influential in setting the nation's dietary menu.[18] In an attempt to reason with woman the nutritionist, Libby's advertisement in the February 1933 issue of *Ladies' Home Journal* announced: "Pineapple for your health's sake." "Have you heard the latest discoveries about canned pineapple?" the text confides. "New food research now shows it to have amazing dietetic values. Eaten daily, canned pineapple contributes to your health in more ways than have been demonstrated for any other fruit!"[19] Such claims of the fruit's nutritional value would continue throughout 1933 and 1934 in the *Journal* by such companies as Dole and Libby, spearheaded by the "educational committee" of the Pineapple Producers Cooperative Association, based in San Francisco.[20] Behind that publicity campaign were the U.S. canned and processed foods industry and its contention that its goods, including pineapples, measured up to and exceeded modern, scientific standards of nutrition and hygiene.

In the midst of popular concern over canned and processed foods, the *Journal* ran a piece, "This New Era in Foods," that reads suspiciously like a conspiracy of magazine publishers, food manufacturers, and advertisers. The article assures its readers that chemists and bacteriologists, "microbe hunters who guard our food supply today," patrol food-processing plants to "guard each step of the manufacturing process from beginning to end." In addition, those scientists proofread their company's advertisements to ensure accuracy in the claims they make. "Canners are among the most assiduous students of science," the writer asserts, having completed "a mountain of research" that proves that canning loses less vitamins than "the usual methods of home cooking." "What does the American woman want?" the author asks. "By and large, I take it she wants leisure, freedom, simplification of household tasks, a full life and a happy one for herself and those dear to her," in effect, the promise of modernity. American women must do their part to secure a full and happy life, "but the food manufacturer does all he can to contribute to the other three." As if in anticipation of reader skepticism over truth in advertising, the author promises that, although profits remain a hallmark of industry, "the trend is set away from cunning toward honesty, toward seeking out and relying on the facts."[21]

Cookbooks, widely believed to have begun in the United States with Amelia Simmons's *American Cookery* (1796), mirrored the period's sentiments about modernity and homemaking. Sarah Tyson Rorer, influential author and founder of the Philadelphia Cooking School (1882–1903), acknowledged in her domestic science text, *Mrs. Rorer's New Cook Book: A Manual of Housekeeping* (1902), that "a great change in the methods of living has taken place in America during the last few years," referencing the modern condition. In the past, she remembered, cooking schools taught cooking only, whereas in the present they must teach "health, body building, and economy in

time and money." She summed up her school's pedagogy: "Cookery puts into practice chemistry, biology, physiology, arithmetic, and establishes an artistic taste. And
if our motto is, 'Let us live well, simply, economically, healthfully and artistically,' we
have embraced all the arts and sciences." Rorer's manual offered cooking tips and information on the chemistry of foods and the latest kitchen gadgets, lists of fruits and
vegetables in season, advice on table waiting, and in its final section Jewish, Spanish,
Creole, and Hawaiian recipes.[22] Another facet of American modernity, the cookbook
became an instrument of advertising, with its mention and use of products, from
kitchen supplies to ingredients, and its turn from farm produce to processed and
canned and later frozen foods to save time and energy.[23]

CONSUMING WOMEN

Modern advertising involved a psychology and aesthetic to seduce the senses and the
mind. Art, especially modernism, offered marketing inducements because its stress
on design and composition over subject matter was rooted in a psychology of per-
ception. As symbolist painter Maurice Denis explained in the 1890s, "It is through
coloured surfaces, through the value of tones, through the harmony of lines, that I
attempt to reach the mind and arouse the emotions."[24] The attraction of advertising
for modern art, thus, was more than a matter of art patronage; it made business sense
insofar as those designs and texts sought to manipulate their readers. Vanity was one
of those appeals, conceded Charles T. Coiner, art director at N.W. Ayer & Son of
Philadelphia, the firm that sent Georgia O'Keeffe to Hawai'i. His commissions to
French artists such as Pierre Roy and Raoul Dufy, he wrote, were negotiated "to reflect
the glamour and atmosphere of the French Line [a steamship company and client],
to express an authentic note of originality and novelty." Those advertisements, Coiner
added, were aimed only at readers of "class publications" who were "by and large, an
informed and civilized group."[25]

The Matson Line, owned principally by the Big Five, including Castle & Cooke,
whose S.S. *Lurline* transported O'Keeffe to the Islands, commissioned modernist
artist Frank Macintosh to create images for its publications during the late 1930s and

'40s. Matson hauled not only passengers between Hawai'i and the West Coast but also sugar and pineapple financed by the Big Five sugar factors. With Big Five patronage and later ownership, Matson held a virtual monopoly, carrying tropical products from the Islands to the continent, and tourists, especially from the 1920s, in the opposite direction. Castle & Cooke and Matson, by building luxury passenger ships and hotels, took the lead in the burgeoning tourist trade.[26] Matson catered to the race and class targeted by N. W. Ayer & Son—the moneyed, informed, and "civilized group."

Exuding taste and elegance and a tropical sumptuousness, Macintosh's designs for Matson were uniquely modern in their simplicity and exemplary in summoning the sensual. "He created a world where slim, langorous women lounged amidst stylized blossoms, epitomizing the dreamy Polynesian rapture that would drift over the visitor upon arrival," wrote an archivist of Macintosh's work for Matson. Although his covers graced Matson menus and brochures for only about a decade, Macintosh's blend of "the 1930's moderne style with a Hawaiian theme" still evokes tourism's images of the Islands, and his renderings remain highly sought collectors' items.[27]

The simplicity of lines, the basic quality of colors, and the representations of life abundant in Macintosh's paintings enlist feelings that have come to mark the modern and its attractions and apprehensions. As a St. Paul, Minnesota, schoolgirl recorded in her notes for homemaking class in 1937: "Art of today must be created today. It must express the life about us. Ours is a complex age. It is much more complex than any previous age. Invention, machinery, industry, science and commerce are characteristic of today. Individuals must have a way of relaxing from this complexity. . . . Thus, we seek to surround ourselves with those things which have the effect of simplicity and which allow us to relax and forget our restlessness." "Modernism is the style of reason," she summarized, "of square, of circle and horizontal line. Good forms and decoration together with good construction will always appeal."[28]

During this period in the United States, the ascent of the mass media, including newspapers and periodicals, gave buoyancy to the advertising industry. Like the advertisers, print journalists discovered "woman" as reader and consumer. Women's columns of the 1890s spread to newspapers and magazines read by an expanding white middle class, and in them women found lessons on household management, news of

FIGURE 36. *Fruit Harvest* (c. 1940) by Frank Macintosh, a Hawai'i cruise line menu cover for ships between the U.S. continent and the Islands. Rudolf Helder, Metropix Inc., Honolulu.

FIGURE 37. *Luau* (c. 1940) by Frank Macintosh, Hawaiian cruise line menu cover. Rudolf Helder, Metropix Inc., Honolulu.

the latest styles and fashions, and appeals to conspicuous consumption. Low in price and lavish in design and illustrations, such publications as the *Saturday Evening Post* and *Ladies' Home Journal* had wide appeal, and as self-appointed guardians of the public taste they were instrumental in the commercializing of gender and the gendering of commerce. Based in Philadelphia, the *Ladies' Home Journal* turned to N. W. Ayer & Son to advertise and promote its sale, and with its success the *Journal* became one of the most popular magazines in which to advertise.[29]

Hawaiian pineapple's appearance in the *Journal,* thus, reached a specific, national audience. As was decided at a 1915 meeting of the Cyrus Curtis company, proprietor of the *Saturday Evening Post* and *Ladies' Home Journal,* the *Journal* would be calibrated for white middle-class women whose annual family income fell between $1,200 and $2,500, and not at those in the higher or lower ranges.[30] Significantly, the target group involved those who were more likely to use pineapple recipes, unlike those in the higher income brackets, who relied upon domestic servants for kitchen duty, or those in the lower, without the means to spend on luxuries. Pineapple's appeal to that class of consumers trafficked on modernity's productions, its consumerism, its racializations and genderings, its reach into field, factory, and home, and its aesthetics, which aroused articulate and inarticulate wants and yearnings. Moreover, the pineapple as food and symbol, as object of desire and conspicuous consumption, was vested in Hawai'i as a tropical paradise, tourist destination, sensuality, and womanly allure.

SIGNIFYING HIERARCHIES

Pineapple's place on the tables of the privileged of Europe long predated the American modern of the late nineteenth and early twentieth centuries. The fruit's royal pedigree, derivative perhaps of Columbus's gift of the tropics to Spain's queen and king, assured its standing as a commodity rare, exotic, unattainable for most, like the Edenic New World, and a sign of extravagant display and use. Moreover, the "king of fruits" was also the "princess of fruits." Thus gendered, it was praised by some European men, like Englishman Richard Ligon, who lived in the West Indies for three years, as "excellent in a superlative degree, for beauty and taste." Penetrating its flesh was "like a

Thiefe, that breaks a beautifull Cabinet, which he would forbear to do, but for the treasure he expects to find within."[31] With its virginity taken for the "treasure" within, the pineapple and in effect its lands, the tropics, had truly become the possession of privileged European men.

In the colony that would become the United States, the English were the arbiters of civility, style, and grace.[32] The English country home, its architecture, furnishings, and gardens, was a mark of success and standing for the elite class of the early eighteenth century. London supplied the furniture, textiles, cutlery and china, silver, and other luxury goods. Among those imports, carved out of and inscribed on stone, silver, wood, and porcelain, was the pineapple, "one of the most prevalent decorative forms of the age in England," according to one account.[33] Stone pineapples on gateposts and a molded pineapple above the front door adorned William Byrd's Virginia home around 1730, and the pineapple appeared above the doorway of the Poynton House in Salem, Massachusetts, in 1750. Pineapple ware, including teapots, mugs and cups, bowls and dishes, and silver pots, was sold throughout colonial America, such as in Francis Wade's Wilmington, New Jersey, shop, which advertised in June 1766 its pineapple teapots, bowls, sugar dishes, mugs, cups, and saucers imported from Liverpool, London, and Ireland for sale "at Philadelphia prices."[34]

Another luxury import was the pineapple fruit from the West Indies, especially the Bahamas, during the eighteenth century. Virginia's governor, the Baron de Botetourt, entertained lavishly and set the standard for the colony's high society beginning in 1768. His average monthly expenditures on food alone soared to about 40 percent of his salary, and among his groceries were pineapples bought in quantities of six to twenty when each fruit cost about a tradesman's day wage. John Murray, the fourth Earl of Dunmore, who erected the Dunmore Pineapple (see chapter 4), succeeded the baron upon his death in 1770.

The governor was not alone in trying to impress his contemporaries with his excess; George Washington and Thomas Jefferson, central figures in the decolonization movement, were eager buyers of pineapples during this time, in the fashion of their colonizers. Washington ordered three dozen pineapples from the West Indies in May 1774, and Jefferson bought pineapples and oranges in 1768 and enjoyed with his family a favorite pineapple pudding:[35]

FIGURE 38. Shirley Plantation, established in 1613, was Virginia's first plantation. Construction of the present mansion began in 1723, and later that century Charles Carter added the three-and-a-half-foot wooden pineapple to the roof's peak. Source: Shirley Plantation, Charles City, Virginia.

Pineapple Pudding

Peel the pineapple, taking care to get all the specks out, and grate it. Take its weight in sugar and half its weight in butter. Rub the butter and sugar to a cream and stir them into the pineapple. Add 5 well-beaten eggs and 1 cup of cream. It may be baked with or without the pastry crust.[36]

The pineapple, centered and rising above a mass of common fruits, conveyed ostentation, wealth, power, and the worldliness of its proprietor with ties not only to mother England but to her daughter colonies like the West Indies in relations of commerce and empire.

Abraham Redwood, son of a plantation owner in Antigua of the Leeward Islands

FIGURE 39. A pineapple as a decorative finial sits atop this Argand wall lamp, which was probably crafted in England or France c. 1790 and was used by George and Martha Washington in their Mount Vernon home. Courtesy of the Mount Vernon Ladies' Association.

in the eastern Caribbean, moved with his family to Newport, Rhode Island, in 1712 as a child. As a man, the immigrant made a fortune by trading New England timber and fish for West Indian sugar and slaves while maintaining the family plantation in Antigua. Two years after his marriage in 1727, Redwood bought two carved pineapples to place atop the gateposts of his home. In the gardens and hothouses of his country home in Portsmouth, tended by a gardener from England, he planted fruits familiar to his island childhood sent to him by his plantation manager in Antigua: oranges, lemons, limes, pineapples, figs, guavas, and tamarind.[37] Redwood embodied a noble man of the English diaspora and empire, with extractive roots in the tropics that enriched and fed the temperate colony and core.

Another man of empire, albeit of a different dominion and age, was John Perkins Cushing of postcolonial Massachusetts. A member of a new nation, the United States, Cushing, like several other prominent Bostonians, amassed his fortune in the China trade. In his famous greenhouses at Watertown, Cushing kept an extraordinary collection of tropical plants, so when two of his pineapples produced fruit in the summer of 1835 the newly launched *American Gardener's Magazine* featured their picking when "perfectly matured." In time, Cushing's hothouses produced significant numbers of pineapples, and his success set off a feverish delirium across Boston's high society.[38] Magazines teased the smitten by taking note of each year's harvest at the Cushing hothouses, like the 1837 crop that featured more than eighty pineapples, many weighing six to eight pounds. We "hope that every gentleman who is fond of fine fruit will possess a pinery," the report recommended.[39]

The *Horticultural Register,* rival of the *American Gardener's Magazine,* agreed that the pineapple was "certainly one of the noblest acquisitions to the Hot House, and forms one of the most prominent features on the table as a dessert." Marshall Pinckney Wilder, president of the Massachusetts Agricultural Society from 1841 to 1848, announced his intention to grow "this greatest of all luxuries" on his Dorchester estate in 1835, and John Lowell's garden in Roxbury produced "this most excellent of all fruits" the following year.[40]

Still, because pineapple cultivation was "very expensive and troublesome," *Family Magazine* noted in 1836, the fruit was seldom grown in the United States and remained, as was reported a decade later, "a production of the tropics," although "one of the most esteemed fruits."[41] In some places, wrote a professional gardener in Albany, New York, pineapple culture was uncomplicated and cheap, but in others, he conceded, it was an "extravagant expense." A hothouse, he offered, moderated those costs and grew the fruit to its "greatest perfection."[42]

Ostentation and conspicuous consumption were uneasy denizens of puritan New England. One of its native sons, Henry David Thoreau, sang the praises of simplicity and frugality, couching those virtues in terms of experience and learning from the local and natural as opposed to the foreign and commercial. Newspapers and politicians, the poet of Walden Pond wrote in 1860, glorify the market success of ships that bring "a load of pineapples" from the West Indies. Yet, he advised, "do not think that the fruits of New England are mean and insignificant, while those of some foreign land are noble and memorable. . . . Our own, whatever they may be, are far more important to us than any others can be. They educate us, and fit us to live in New England.

Better for us is the wild strawberry than the pineapple. . . . the wild apple than the orange, the hazelnut or pignut than the cocoanut or almond, and not on account of their flavor merely, but the part they play in our education."[43]

Unlike their northern brethren many wealthy planters in the South paid little notice to English gardens, yet sectional differences, the Civil War notwithstanding, failed to sour their and America's taste for imported luxuries like the pineapple. In the war's midst, the Bahamas exported nearly 700,000 pineapples to the United States, and some of those ships slipped past Union blockades to deliver tropical fruits—pineapples, bananas, and oranges—to the Confederacy.[44] After the war, with New York City as the main port of entry, nine-tenths of the fresh pineapple imports, an 1877 *Philadelphia Inquirer* article estimated, arrived from the Bahamas, and in 1881 an American traveler reported that those islands produced significant numbers of canned pineapples.[45]

Florida, starting in the 1870s and reaching a peak in 1910, produced pineapples, supplying New York and other northern markets, although frosts and poor soils presented challenges not faced by tropical island growers.[46] Notable among that state's pineapple producers of the early twentieth century were Japanese immigrants, who in 1904 founded the Yamato colony of south Florida, which, like its relative in California, was the progeny of idealists and intellectuals in pursuit of their version of the American dream during a period of "yellow peril" clamor against the Japanese of the "Yamato race" and Asian exclusion and expulsion accomplished by U.S. statutes.[47] By the close of the nineteenth century, Florida's pineapples and West Indian imports had reduced the fruit's price, making the once rare luxury "as common as native products."[48]

Yet American housekeepers required instruction in the management of the alien pineapple, as seen in many magazine articles on "how to eat . . . pineapples" and "the best and easiest way" to handle and prepare that "welcome visitant from tropic climes." In an 1875 notice, James W. Parkinson from Philadelphia warned his readers against cutting the pineapple with a knife, when "tearing it to pieces" was the "natural" method of preparation. "You damage both its flavor and its healthfulness," Parkinson explained, by cleanly cutting the fruit. To commence "the process and pleasures of eating," "chip off" the "bark," scoop out the "eyes," and press a fork at a forty-five-degree angle into the flesh, tearing off a "large bite." "Crush one of these large hunks between your teeth," he relished, and that "luscious and highly flavored pineapple juice will delight your palate and gurgle rejoicingly down your throat," leaving you with a taste never before imagined possible.[49] Uncivil consumption had its joys.

In addition to its taste divine, family managers of the late nineteenth and early twentieth centuries were informed, pineapple possessed "wonderful medicinal qualities as a digestant, and is highly prized by up-to-date physicians, in treating gastritis and other serious affections of this nature."[50] Another report proclaimed the "medicinal virtues of the pineapple." Fresh pineapple juice, its author held, "contains a remarkably active digestive principle similar to pepsin. This principle has been termed 'bromelin,' and so powerful is its action upon proteids that it will digest as much as 1,000 times its weight within a few hours." An expert testified that the U.S. Dispensary routinely recommended pineapple juice "for gastric indigestion and [it] is used by the laity as a gargle for sore throats. It is quite possible," the professor and food editor speculated, "that its enzyms have some therapeutic properties."[51]

Pineapple modern, thus, gained admittance into the American domestic sphere as food and medicine, and also as design. Crochet is a way of creating fabric from a length of thread using a crochet hook.[52] A chain of loops forming patterns, crochet substituted for more expensive kinds of lace in nineteenth-century Britain, France, and the United States. As the price of cotton thread dropped, crocheted laces supported a thriving cottage industry among working-class women, for sale to the burgeoning middle class as a status symbol. Crocheted lace, although initially labeled a cheap imitation, became more elaborate and complicated in stitching and design during the modern period. Although first produced by hunters and fishermen,[53] thread work and crocheting came to occupy women's hands during "leisure" hours in the home, allegedly to insulate them from evil. This gendered division of labor also mirrored the assumed distance between mind and body, head and hand, whereby men arranged the patterns for the fabrics and women, originally cast by men as incapable of creative thought, simply executed the contriver's compositions.[54]

There were women, against those odds, who rose to design their own fabrics, and some came to realize the slender freedoms conferred to them by their earnings in needlework. Under modernity, handmade objects such as women's needlework signified another kind of escape from the clutches of the machine and uniformity. And as with industry's migration from Europe to America, European needlework and its revivals influenced those labors in the United States by way of exhibits, such as the display by the Royal School of Art Needlework at the Philadelphia Centennial of 1876 and many pattern and technique books published in the "Old World." Crocheting may have also migrated to the United States with the Irish, who in the nineteenth century were renowned for their special form, Irish crochet, and thriving cottage industry.[55] Mid-nineteenth-century immigration, one history recounts, fostered the early development of crochet in America, and that beginning reveals "an obvious Irish influence with the style of rosettes, leaves, picots and other motifs."[56]

One of the most popular crochet patterns was the pineapple, which may have originated in the ancient Greek veneration of the pine tree as a source of renewed energy and strength—modified among the Turks and Arabs as a symbol of life and life everlasting.[57] The slippage between pine and pineapple as design recalls the move from the Spanish *piña* to the English "pine apple," and readers of U.S. magazines of the late nineteenth century found patterns for pineapple lace work, said to resemble "the divisions of the pineapple or the scales of the pine cone."[58] Although it is unclear how or when the pine or pineapple motif and its probable meaning in needlework migrated from Europe to America, depictions of nature in fabric commonly held symbolic significances of life and growth, beauty and abundance. Embroidered renderings of the pomegranate, palm, lotus, and pineapple from Turkey, Arabia, Persia, India, China, and Japan, an American needlework editor explained in 1881, reflect "the patient exact toil which produces these Oriental embroideries, whose workmanship is without flaw, [and they] can only be found in those countries of the sun where the days are longer than ours."[59] Geography ("countries of the sun") and human works ("Oriental embroideries") form correspondences in this explanation.

"Our fashionable ladies," an 1879 article on lace making declared, were involved in "art needlework," although lace schools were few in number, and American laces were "crude" when compared with those of "the finest lace artists of Europe."[60] A year later, a report on an "art embroidery revival" conceded that American women followed "our British sisters" but added, "we have often improved on our models."[61] In 1885, Laura Safford helped to launch the Philadelphia Needlework Guild, which evolved into the Needlework Guild of America. Publications, advertising, and marketing popularized crocheting as a craft and hobby.[62] At the start of the twentieth

FIGURE 40. Pineapple lace. *Arthur's Home Magazine* 61 (August 1891): 657.

century, however, machine-made laces led to a decline in crocheting, and World War I, "the great war" and fratricidal conflagration, marked an end to the exuberance of empire and its rivalries but also of solidarities of race,[63] along with a disillusionment with the grandeur of old European monarchies and their fashions of rich embroideries and lace.[64]

Allied to these images of distinction and ostentation are their associations with fashion and style, domesticity, and possibly hospitality. "Because the fruit played such an important part in the social life of the time and appeared as a decorative motif on so many objects which had to do with the welcoming, sheltering, entertainment, and refreshment of guests," a writer speculated in 1945, "the idea of hospitality became attached to it and is now fairly widespread."[65] A more recent version attributes to the Colonial Revival movement the selection of the pineapple as a symbol of welcome, a fiction that gained currency in the 1930s, when historic houses and museums, many with pineapple gateposts and dinnerware, concocted that symbolism to recuperate an idealized and romanticized colonial past. Ship captains, the yarn goes, skewered pineapples on their wrought-iron gates to indicate their return from the tropics and to invite their friends and neighbors to join in the welcome and celebration.[66]

The so-called Colonial Revival began mainly in New England in the 1870s during the nation's first centennial, and it celebrated house exterior and interior designs and furnishings as artifacts of a vanishing, golden age. During the early 1900s, the Metropolitan Museum of Art displayed actual rooms from colonial homes shipped to its New York City galleries, and in 1932 landscape architects commemorated George Washington's two-hundredth birthday by featuring colonial gardens. Washington's Mount Vernon and Jefferson's Monticello and other historic homes became popular tourist destinations; restoring colonial homes and adding colonial features to venerable Victorian homes became fashionable.

Some cultural historians point out that those retrospectives gained ground during the late nineteenth and early twentieth centuries, a period of enormous and rapid changes that typified modernity but also a time of unprecedented immigration from southern and eastern Europe, Asia, and Latin America and internal migration of African Americans from the South to the North, all of which transformed the face of urban, industrial America.[67] An anchor in a surging sea of changes, the "colonial," defined as Georgian and Neoclassical styles of the eighteenth century, promoted ideas of democracy, patriotism, taste, and moral superiority through architecture, landscape and garden design, and the decorative arts.[68] Those core virtues reinstated the centrality of "white," manly America, absent darker and weaker Europeans, peoples of color, and a "babble" of languages, religions, and cultures, even as the nation was engaged in the acquisition of a tropical empire in the Caribbean and Pacific.[69]

PINEAPPLE CULTURE

The pineapple as a fruit of the tropics and a trophy of empire was but one of numerous material and symbolic objects of desire that prompted movements across the temperate and tropical zones in world history. Fueled by wants, those transgressions of places produced mappings to chart trails but also to name, describe, and classify novel airs, waters, sites, and peoples. The ancient Greeks theorized oppositions of inhuman heat—the tropics—and cold—the polar extremities—and the mediation of a hospitable temperate band. Consorts of temperature were tropical wetness, signifying woman, and temperate dryness, signifying man. Climates and constitutions, racialized and gendered, formed coterminous, stable states wherein the environment determined the natures and rankings of their natives. Later, actual crossings of those hypothesized divides, notably Alexander the Great's invasive thrust into Asia and engagements between the Roman and Arab worlds, expanded considerably and modified European imperial mappings of their "world island" even as they affirmed certain theoretical projections despite the rub of contact and experience.

Persistent was the notion that environments predisposed qualities and its ally, the idea that those characteristics were "in the blood" and transmitted from one generation to the next, giving rise to gendered "races." Exploration, trade, conquest, colonization, and empire across race/gender isotherms carried self-described "superior," manly natives of the temperate zone to the tropics, bringing to the "inferior," womanly natives of the torrid zone the gifts of civilization and Christianity, to some, for the betterment of all and to others, the stirring of inert, feminized races and their recovery and movements that threatened to overrun the white man's turf. And whether at home or on the fringes of empire, interactions among gendered races promised to pollute allegedly pure bloodlines and introduce diseases like yellow fever, which broke out in Philadelphia in 1793. Attributed to Santo Domingo's political refugees, the alien contagion revealed the intimate ties of commerce and empire (Pennsylvania's soldiers having served in the West Indies during the 1740s) that attached the tropical to the temperate zone and the dangers they posed even as they delivered enormous profits to the city's merchants from the traffic in sugar and slaves.

Plantations were generally installed by conquerors with migrant laborers, because of horrific declines among the native peoples, to produce a single, export crop. That form of production plied the oceans—the Mediterranean, Atlantic, Caribbean, Indian, and Pacific—with the global spread of Europeans. Millions of native people died and were displaced as a consequence, and millions of enslaved and indentured Africans and Asians were transported to and labored in plantations on islands and continents. Conquest, a U.S. president confessed before falling into the deep sleep of the blessed, was divinely ordained "to educate the Filipinos, and uplift and civilize and Christianize them," while others slept fitfully, worrying over the residues of empire, the "rising tide of color" that menaced the inner dikes. Those reciprocals of peril and profit charted the paths of commerce and empire, which can be traced through the peregrinations of the pineapple, the "princess of fruits."

As indigenous as the proto-Tupí Indians who found, selected, and bred it, the *nana* was, like the native peoples, a migrant and a creation. Carried by the descendants of its discoverers and others from the interior to America's coasts, the pineapple was a rare and exotic food and plant that grew, like its handlers, indigenous to its new homes along Ecuador's lowland coast, the upper reaches of the Orinoco River, Panama's coast, and the islands of the Caribbean, where on Guadeloupe the Callinago introduced the fruit to Columbus. The admiral brought on board the pineapple, a captive of that "discovery" like his so-called Indians, to present to his sponsors, Spain's queen and king. Whereas others of America's gifts to the world, such as the potato, corn, beans, and sweet potatoes, were far more influential in directing the course of human events, the pineapple acquired a noteworthy status as produce and symbol.

Cultivated in Europe's gardens—enchanted isles—and hothouses by the elite class, the pineapple was a prize of plunder, a taste of the Edenic "New World," and an object rare, coming to fruit only in assimilating environments that reminded it of its distant, tropical home. Exhibited on tables of excess, the fruit, likened by some men to a comely, virginal woman, was contested over as an article of personal and at times national rivalries, mirroring the larger conflict over possessions of land, labor, and goods. The science of acclimatization was another conquest of the temperate over the tropical world, and it accompanied and advanced European empires whose strategic networks included gardens in the core and selected colonies and agents and collectors at the peripheries, constituting an empire of plants.[70]

The pineapple, from its strange place of capture and exile, like the Cayenne variety, remigrated to more familiar climes in Australia, Florida, Jamaica, and Hawai'i, and from those Pacific islands the Cayenne returned to Caribbean America and spread to Southeast and East Asia, Indian Ocean Africa, and other Pacific islands. Those Cayenne crossings and destinations were plotted to produce on vast plantations, on cheap yet fertile ground with inexpensive and efficient labor, fresh and processed fruit for foreign, more lucrative markets and gentler palates.

Capitalism's markets, like the clipped, cored, trimmed, and sliced wild fruit, were to a large extent fabrications of human desires and needs. A man of his times, James Dole understood and practiced that economic maxim well. Son of a clergyman and a member of the "mission boys" in Hawai'i, Dole followed the precept of his elite band, involving a spiritual kingdom on earth of plantations and factories ostensibly to convert Hawaiian natives and colored migrants, uncivilized heathens both, to Europe's "holiest possession," Christian civilization. Instead, the mission boys found a new faith in the Islands to sanctify their overthrow of the Hawaiian kingdom and worldly seizure of the political, economic, and social life of the republic and then territory. With king sugar and his consort, princess pineapple, holding the reigns, Dole's Hawaiian Pineapple Company and its Big Five proprietor, Castle & Cooke, followed in modernity's tracks to infiltrate the hearts, minds, and homes of America's mainly white middle-class women.

Modernity enabled the pineapple's move from a luxury attainable only by the rich to a canned convenience for the growing middle-class masses, while a nostalgia for the simplicity of the colonial era materialized into a style and movement amidst rapid and far-reaching transformations of demography, industry, and empire, which unsettled a founding myth of homogeneity, exceptionalism, and a uniform descent and tradition. Manufactures such as the pineapple as an object of desire and status symbol were conceits of the core as well as the periphery from whence came Dole's choice advertising to stack every supermarket shelf and his innovation of identifying a product with a place and sentiment—the "Hawaiian pineapple," his contributions to modern America.

Bearing the Hawaiian label, the pineapple acquired its significations of tropical, womanly paradise, fecundity, health, and bodily pleasures and freedoms, without irony a flight from muscular modernity. With tourist dollars to be had for luxury lines like Matson and hotel shareholders like Castle & Cooke after the decline of sugar and pineapple plantations in the Islands, the pineapple as a symbol of hospitality and generosity, a pining for an imagined New England past, was made local by the tourist industry as the "aloha spirit," which mocked Hawaiian dispossession and leied and embraced tourists in advertisements, brochures, and postcards.[71] Since 1989, the "Pineapple Experience" at the Dole Plantation near the founder's original Wahiawā homestead has become a major tourist destination, welcoming more than one million visitors each year.

Agricultural science, a correspondent for the *New-England Farmer* wrote in 1852, was beginning to dawn in "the isles of the Pacific" with the cultivating of such farmers from New England as William L. Lee, a native of western New York, chief justice of the supreme court of Hawai'i, and a founder of the Royal Hawaiian Agricultural Society. At its preliminary meeting held in 1850, Lee was elected president, and the Society began the business of advancing agricultural production. That development was profound, the "interested" and "amused" reporter noted, in the light of history. "Fifty years ago those islands were peopled with countless multitudes of naked savages, of the lowest grade; ignorant alike of the culture of the earth and of the God who made it. The Anglo-Saxon came among them; and though the native population has dwindled to a comparative handful, it has become Christian. . . . native nakedness is clothed; the wilderness has been made to blossom as the rose, and the earth to pour forth her abundance."[72]

Such sentiments were entirely without merit. Hawaiians were highly skilled horticulturalists, among the most advanced of all of Oceania's peoples, and they produced a prodigious variety and quantity of foods especially suited for their island world.[73] Individuals such as Archibald Menzies, who visited Hawai'i as a member of an expedition led by British captain George Vancouver in 1793, testified to the scale of those works. He described breadfruit plantations on the island of Hawai'i: "The space be-

FIGURE 41. The Dole Plantation, a tourist destination. Gary Y. Okihiro photograph, 2008.

tween the trees did not lay idle. It was chiefly planted with sweet potatoes and rows of cloth plant. As we advanced beyond the bread-fruit plantations, the country became more and more fertile, being in a high state of cultivation. For several miles round us there was not a spot that would admit of it but what was with great labor and industry cleared of the loose stones and planted with esculent roots or some useful vegetable or other."[74] The evidence, one scholar concluded, shows that Hawaiians "had an extraordinarily intimate and thorough knowledge of the many varieties of taro, sweet potato, sugar cane, and banana . . . , and that, in their selection of plants and methods of cultivation, they practiced what I believe most agriculturalists would agree was definitely an advanced art of gardening."[75]

Writing against those realities, colonial histories record the gift of civilization and progress bestowed upon the impoverished tropical margins in violation of and triumph over alleged spatial natures. European men deployed science in that act of benevolence to reproduce the tropics in the temperate band even as white skins adapted to the torrid zone sun's fierce rays and white bodies conquered its diseases. Science, thereby, enabled and built empires by bringing rational economic and botanical order to lands formerly thought of as white men's graves trampled by beasts, Amazons, and perversions of nature.[76] And in the metropole, industry improved upon the original. An 1884 essay on tropical fruits affirmed, "It is the fashion to speak of our English hot-house fruit as superior to any tropical product."[77] The problem, authorities agreed, lay with the natives of the tropics, who had little concern over the fertility of the soil or the careful selection and improvement of plants; instead, they were contented and lived "a life of idleness with little thought for the morrow." The lush, womanly tropics, a place of plenty, was like its natives, inert, undisciplined, and "without the systematic application of scientific principles." Therefore, "part of the 'white man's burden' has been in the tropics to revolutionize agriculture."[78]

The modern empire, a report on the "new agriculture of the tropics" noted, held the key to a bright future, as the English, Dutch, Germans, and Americans had shown in their tropical possessions. "Before white men settled in tropical America," the writer recounted, "the sugar industry was in the most primitive condition. Machinery for extracting the juice of the cane was unknown, and the plants were semi-wild growths that yielded a very small percentage of sugar." Whites introduced science to that wilderness and conquered it, adding "hundreds of millions of dollars of wealth to tropical America" and giving "regular employment to the natives." The prospects were monumental. "We are just on the threshold of developing the world's crop of fruits," the article rhapsodized, as had visionaries Andrew Preston and James Dole. "From South and Central America, from the islands of the Pacific and Atlantic, from equatorial Africa, and from the lands of the Orient, streams of tropical fruits will in the near future pour into Europe and America in return for the cereals, meats, and products of the colder climes. Under modern agricultural methods, an abundance of fruits for the whole world can be raised in these warm regions at a cost so low that none need be so poor as to go without them."[79]

But development of the core was purchased with the underdevelopment of the periphery, with deadly consequences for native peoples. They were "devoured," in the prophetic words of Hawaiian Davida Malo. Alien intruders—germs, ideas, material objects, and "clever men from the big countries"—decimated the population, undermined subsistence production and religious beliefs, engineered land losses and therewith the people's mana, transformed the mode and relations of production, and diminished and then extinguished the kingdom's sovereignty. Impostors, the last Hawaiian ruler of the Islands charged, schemed "to defraud an aboriginal people of their birthrights." Those foreign machinations and implantations were sustained by imported laborers who as docile, efficient workers served their masters but as agents claiming redress and rights were registered as "perils" and the vanguards of a global race war. Specters, thereby, haunted the dreams of abundance and fabulous returns.

Tropical fruits—like tropical climates, lands and waters, plants and animals, gendered races and civilizations, and diseases—were alien and deviant when set in opposition to the European familiar and normal and its temperate airs, waters, and sites. Upon entering the "equinoctial regions of America," Alexander von Humboldt observed in his tropical hermeneutics, the southern skies appeared totally unfamiliar, constituting an "unknown firmament" in another world. These hypothesized antipodes operate to define the superior self even as they vilify the inferior other. For naturalist Henry David Thoreau, imported fruit from the distant tropics only whetted his desire for the simple pleasures of homegrown New England produce. That preference, while at variance with some of Thoreau's own values, underscored regional and national scripts of loyalties to place, kin, and country, but it also consorted with the period's transnational identities of white and colored, male and female, colonizer and colonized, civilization/Christianity and barbarism/paganism, and the temperate and tropical bands.

Despite these manufactured and measured differences and alienations, both "worlds" were integrated, not in the nature of convection currents but in the human enactments of travel, migration, and empire, from mercantile and then colonial (im)plantations and their circulations of capital, labor, goods, and culture.[80] These movements, suggestive of space, time, and agency, were facilitated by scientific expeditions and botanical gardens and collectors, constituents of an imperial science and discourse inhabited by writers, geographers, historians, and "naturalists" authorized to name, classify, and exert dominion over lands and waters, climates, plants, animals, gendered races, and racialized contagions at the fringes and within the heartland in a process of assimilation, of consumption.

As was feared by some, the infiltration of the tropics spread to virtually every home and homemaker in the United States, an invasion attenuated and packaged by advertising and a domestic science of nutrition and convenience appealing to the "inarticulate longings" of white middle-class women. The pineapple, in addition to its access to the domestic sphere as food, graced fenceposts and entryways, serving utensils and vessels, and crocheted patterns on tabletops as design and symbol. And acclimatization, like assimilation, domesticated alien organisms, rendering them safe for the homefront even as Hawai'i's annexation and the absorption of other distant colonies made their products "domestic," contained through that gendering and made compassionate by modernity and its interventions.

Pineapple culture exemplifies these propositions and productions. Shorn of its wildness, the "manhandled," mass-produced pineapple appears in its natural state, sized and boxed for the traditionalist and in its clipped, dissected, and sweetened state for the modernist. The fruit reaches both markets and tastes year-round in disregard of seasons and places and advances a global, material culture and signification that qualifies and diminishes the contrived distances between the polarities of West and East, continents and islands, temperate and tropical zones.

NOTES

INTRODUCTION

1. *Pineapple Culture* is not a history of the fruit or of the fruit as food. Fran Beauman, *The Pineapple: King of Fruits* (London: Chatto and Windus, 2005), is an able history of the fruit. For a recent, sweeping history of food, see Paul Freedman, ed., *Food: A History of Taste* (Berkeley: University of California Press, 2007).

1. MAPPING DESIRES

1. H. F. Tozer, *A History of Ancient Geography,* 2d ed. (London: Cambridge University Press, 1935), 12.

2. J. Oliver Thomson, *History of Ancient Geography* (Cambridge: Cambridge University Press, 1948), 94–95, 97.

3. Some attribute to Parmenides, who lived about a generation after Pythagoras (c. 569–475 B.C.), this notion of zones. E. H. Bunbury, *A History of Ancient Geography,* vol. 1, 2d ed. (New York: Dover, 1959), 125; and Tozer, *History,* 60.

4. *Hippocrates,* vol. 1, trans. W. H. S. Jones (Cambridge, Mass.: Harvard University Press, 1923), 105, 107, 109, 113, 115. See also Clarence J. Glacken, *Traces on the Rhodian Shore: Nature and Culture in Western Thought from Ancient Times to the End of the Eighteenth Century* (Berkeley: University of California Press, 1967), 80–115. Asia's moisture, heat, and sensuality parallel women's wetness, softness, gluttony, and insatiable sexuality in these representations. See Anne Carson, "Putting Her in Her Place: Woman, Dirt, and Desire," in *Before Sexuality: The Construction of Erotic Experience in the Ancient Greek World,* ed. David M. Halperin, John J. Winkler, and Froma I. Zeitlin (Princeton, N.J.: Princeton University Press, 1990), 137–45; Leslie Dean-Jones, *Women's Bodies in Classical Greek Science* (Oxford: Clarendon Press, 1994), 41–109, 115, 120–21, 123–

25; and Helen King, *Hippocrates' Woman: Reading the Female Body in Ancient Greece* (London: Routledge, 1998).

5. *Hippocrates,* 133–35, 137.

6. Bunbury, *History,* 395–97; and Tozer, *History,* 179.

7. Aristotle, *The Politics,* trans. Benjamin Jowett (Cambridge: Cambridge University Press, 1988), 74, 165.

8. Richard Stoneman, "Introduction," in *The Greek Alexander Romance,* trans. Richard Stoneman (London: Penguin Books, 1991), 3.

9. Thomson, *History,* 134–35.

10. On the Rome and India trade, see Vimala Begley and Richard Daniel De Puma, eds., *Rome and India: The Ancient Sea Trade* (Madison: University of Wisconsin Press, 1991).

11. Donald F. Lach, *Asia in the Making of Europe,* vol. 1, *The Century of Discovery* (Chicago: University of Chicago Press, 1965), 14–16, 19, 20, 23.

12. Thomson, *History,* 321, 324, 328, 389.

13. On Bodin's standing in the history of environmental theory, see Glacken, *Traces,* 434–47.

14. John Bodin, *Method for the Easy Comprehension of History,* trans. Beatrice Reynolds (New York: Columbia University Press, 1945), ix, xi–xiii.

15. For a discussion of Plato's views on "race," see Ivan Hannaford, *Race: The History of an Idea in the West* (Washington, D.C.: Woodrow Wilson Center Press, 1996), 30–43.

16. Travelers, including Marco Polo, traders, and missionaries had in the thirteenth and fourteenth centuries dispelled the idea of an uninhabitable tropical band. Lach, *Asia,* 32, 35–36, 43, 47–48.

17. Bodin, *Method,* 87, 92–96, 116–19.

18. Bodin, *Method,* 86, 97, 101, 102, 103.

19. Hannaford, *Race,* 156.

20. Johann Friedrich Blumenbach, *On the Natural Varieties of Mankind,* ed. Thomas Bendyshe (New York: Bergman, 1969), 71–81, 98, 99.

21. Blumenbach, *On the Natural,* 100, 101, 102, 104, 110, 111, 113.

22. Blumenbach, *On the Natural,* 188, 192, 193, 196–203, 207, 209, 210, 211–12, 215; and Bodin, *Method,* 102, 107.

23. Blumenbach, *On the Natural,* 223–57.

24. Blumenbach, *On the Natural*, 269; and Hannaford, *Race*, 207, 208.

25. Bodin, *Method*, 143.

26. Blumenbach, *On the Natural*, 224.

27. Blumenbach, *On the Natural*, 275, 276; and Hannaford, *Race*, 211, 212.

28. Arthur de Gobineau, *The Inequality of Human Races*, trans. Adrian Collins (New York: Howard Fertig, 1967), 24, 25, 27, 28, 31. See Mark Harrison, *Climates and Constitutions: Health, Race, Environment and British Imperialism in India, 1600–1850* (New Delhi, India: Oxford University Press, 1999), for an affirmation of this fear in a colony. For Harvard geology professor Louis Agassiz's tropical researches (1865–66) to affirm racial degeneration among Brazil's "racial hybrids," see Nancy Leys Stepan, *Picturing Tropical Nature* (Ithaca, N.Y.: Cornell University Press, 2001), 85–119. Cf. Stephen Jay Gould, *The Mismeasure of Man* (New York: W. W. Norton, 1981), 42–50, on Agassiz's encounters with blacks in the United States and another origin for his fears of miscegenation.

29. At the same time, Princeton geographer Arnold Guyot remained convinced of climate's sway over human energies and capacities. Arnold Henry Guyot, *The Earth and Man: Lectures on Comparative Physical Geography in Its Relation to the History of Mankind* (New York: Sheldon, Blakeman, 1849). See also Josiah Clark Nott and George R. Gliddon, *Indigenous Races of the Earth; Or, New Chapters of Ethnological Inquiry* (Philadelphia: J.B. Lippincott, 1857). On the prevalence of that view among nineteenth-century scientists, see David N. Livingstone, "The Moral Discourse of Climate: Historical Considerations on Race, Place and Virtue," *Journal of Historical Geography* 17:4 (1991): 413–34.

30. Gobineau, *Inequality*, 36–62. Okihiro, *Island World: A History of Hawai'i and the United States* (Berkeley: University of California Press, 2008), the first volume of my trilogy, discusses Christian missions and their ostensible uplift of Hawaiians.

31. Gobineau, *Inequality*, 116, 146, 150–51, 154–67, 179.

32. Ellen Churchill Semple, *Influences of Geographic Environment: On the Basis of Ratzel's System of Anthropo-Geography* (New York: Holt, Rinehart and Winston, 1911), 10.

33. For a study of Rätzel's role in advancing the imperial German state, see Mark Bassin, "Imperialism and the Nation State in Friedrich Ratzel's Political Geography," *Progress in Human Geography* 11:4 (1987): 473–95.

34. See, e.g., Edwin Black, *War against the Weak: Eugenics and America's Campaign to Create a Master Race* (New York: Four Walls Eight Windows, 2003).

35. Semple, *Influences*, 1, 2, 10, 11.

36. Semple, *Influences*, 33–50, 112–13.

37. Semple, *Influences*, 168–70, 183–98. In the colony, elites cited indigenous cultures, race mixtures, and whitening in forming a Brazilian national character and history. Stepan, *Picturing Tropical Nature*, 120–48.

38. Semple, *Influences*, 173–74, 176–77, 181–82.

39. Semple, *Influences*, 607, 608–11, 615–16. See also Robert De Courcy Ward, *Climate, Considered Especially in Relation to Man* (New York: G. P. Putnam's Sons, 1908).

40. Semple, *Influences*, 620, 626–27.

41. Semple, *Influences*, 633–35.

42. Semple, *Influences*, 616, 628.

43. Semple, *Influences*, 616.

44. Semple, *Influences*, 625–26.

45. Ellsworth Huntington, *Civilization and Climate* (New Haven, Conn.: Yale University Press, 1915), 1, 2, 3, 6, 9. See also Ellsworth Huntington, "Environment and Racial Character," in *Organic Adaptation to Environment*, ed. Malcolm Rutherford Thorpe (New Haven, Conn.: Yale University Press, 1924), 281–99; and *The Character of Races: As Influenced by Physical Environment, Natural Selection and Historical Development* (New York: Charles Scribner's Sons, 1924).

46. Begun after the Civil War by Samuel Chapman Armstrong, Hawaiʻi-born son of missionary parents, the Hampton Institute employed manual and industrial education for the "uplift" of African Americans and American Indians. The institute is a subject of Okihiro, *Island World,* chapter 4.

47. Huntington, *Civilization,* 9, 11–12, 16–17.

48. S. C. A. [Samuel Chapman Armstrong], "Reminiscences," in *Richard Armstrong: America, Hawaii* (Hampton, Va.: Normal School Steam Press, 1887), 84.

49. See also Huntington, "Environment and Racial Character," 292, where he calls the people of the tropics "the children of the human race" and representatives of "our primitive ancestors. Their characteristics are those which unspecialized man first showed when he separated from the apes and came down from the trees."

50. Huntington, *Civilization,* 33, 35, 37, 38–40, 42, 43, 44, 110, 139, 148, 271, 286. Huntington's methods and conclusions were hailed and condemned by his contemporaries. See, e.g., David N. Livingstone, "Climate's Moral Economy: Science, Race and

Place in Post-Darwinian British and American Geography," in *Geography and Empire*, ed. Anne Godlewska and Neil Smith (Oxford, England: Blackwell, 1994), 143–44.

51. Charles H. Pearson, *National Life and Character* (London, 1893), 13–14, 32–47, 67–68, 88–90.

52. Benjamin Kidd, *Social Evolution* (New York: Macmillan, 1894), vii, 1–3, 326–29, 330–40.

53. Benjamin Kidd, *The Control of the Tropics* (New York: Macmillan, 1898), 17.

54. James Bryce, "The Migrations of the Races of Men Considered Historically," *Eclectic Magazine of Foreign Literature* 56:3 (1892): 290, 292.

55. James Bryce, "British Experience in the Government of Colonies," *Century Illustrated Magazine* 62:5 (1899): 718–19.

56. See a brief review of the discussion of climate and colonies and U.S. possessions in Robert De Courcy Ward, "Notes on Climatology," *Journal of the American Geographical Society of New York* 31:1 (1899): 160–62.

57. Harrison, *Climates and Constitutions,* reveals that the nature of the colonial project also affected British perceptions about India's climate vis-à-vis white bodies. In the eighteenth century, when the British project involved mainly coastal trade, whites held optimistic views about acclimatization and settlement, but in the nineteenth century, when it involved occupation and rule, they believed that India's climate was incompatible with white constitutions.

58. J.W. Gregory, *The Menace of Colour: A Study of the Difficulties Due to the Association of White & Coloured Races, with an Account of Measures Proposed for Their Solution, & Special Reference to White Colonization in the Tropics* (London: Seeley, Service, 1925), 173–74.

59. Gregory, *Menace of Colour,* 25–30, 173–215.

60. Gregory, *Menace of Colour,* 238–39, 240, 241.

61. Lucien Febvre, *A Geographical Introduction to History,* trans. E.G. Mountford and J.H. Paxton (New York: Alfred A. Knopf, 1925), 1–8, 18, 90.

62. Febvre, *Geographical Introduction,* 91, 98–114, 122–31, 137–69, 171–294, 295–315, 367.

63. Most scholars agree that race as we now know it is mainly a modern idea, although its roots, which may not resemble the present variety, reach farther back in time. See, e.g., Hannaford, *Race;* and Thomas F. Gossett, *Race: The History of an Idea in America* (New York: Oxford University Press, 1997).

2. EMPIRE'S TROPICS

1. Paul Carter, *The Road to Botany Bay: An Essay in Spatial History* (London: Faber and Faber, 1987), xvi.

2. As quoted in Livingstone, "Climate's Moral Economy," 134, 135. See also Brian Hudson, "The New Geography and the New Imperialism: 1870–1918," *Antipode* 9:2 (1977): 12–19.

3. On the erasure and reinscription of empire as "a quintessentially geographical project," see Anne Godlewska and Neil Smith, eds., *Geography and Empire* (Oxford, England: Blackwell, 1994), especially their introduction to the volume.

4. As such, European accounts of America were more a discovery of the European self than of its American other; the "Old World" familiar, including ideas of tropical lands and their flora and fauna, helped to describe and explain "discoveries" in the "New World." See, e.g., J. H. Elliott, *The Old World and the New, 1492–1650* (Cambridge: Cambridge University Press, 1970). More complex, Europe's assimilation of America was layered and multiple, both self-referential and troubling of old truths. See, e.g., Stephen Greenblatt, *Marvelous Possessions: The Wonder of the New World* (Chicago: University of Chicago Press, 1991); and Karen Ordahl Kupperman, ed., *America in European Consciousness, 1493–1750* (Chapel Hill: University of North Carolina Press, 1995).

5. For a study of Columbus's creations, see Valerie I. J. Flint, *The Imaginative Landscape of Christopher Columbus* (Princeton, N.J.: Princeton University Press, 1992). In his imagined world, Columbus was not alone. See, e.g., John L. Allen, "Lands of Myth, Waters of Wonder: The Place of the Imagination in the History of Geographical Exploration," in *Geographies of the Mind: Essays in Historical Geosophy,* ed. David Lowenthal and Martyn J. Bowden (New York: Oxford University Press, 1976), 41–61; and Anthony Pagden, *European Encounters with the New World* (New Haven, Conn.: Yale University Press, 1993).

6. David N. Livingstone, "Tropical Hermeneutics: Fragments for a Historical Narrative, an Afterword," *Singapore Journal of Tropical Geography* 21:1 (2000): 92–98. See also David Arnold, *The Tropics and the Traveling Gaze: India, Landscape, and Science, 1800–1856* (Seattle: University of Washington Press, 2006).

7. On science and Spanish colonialism and national development, see Juan Pimentel, "The Iberian Vision: Science and Empire in the Framework of a Universal Monarchy, 1500–1800," in *Nature and Empire: Science and the Colonial Enterprise,* ed. Roy MacLeod, Special Issue, *Osiris,* 2d ser., 15 (2000): 17–30.

8. Mary B. Campbell, *The Witness and the Other World: Exotic European Travel Writing, 400–1600* (Ithaca, N.Y.: Cornell University Press, 1988), 171, 247. On medieval Europe's location of Paradise in Asia and its sighting by Columbus in America, see Flint, *Imaginative Landscape,* 9, 10–14, 31–32, 102–4, 149–81; and on Columbus's feminization of America as lack or deficiency, timidity, fertility, and object of desire, see Margarita Zamora, "Abreast of Columbus: Gender and Discovery," *Cultural Critique* 17 (Winter 1990–91): 130–33, 134, 137, 143–49.

9. *The Log of Christopher Columbus,* trans. Robert H. Fuson (Camden, Me.: International Marine, 1987), 75–76. Anne McClintock, *Imperial Leather: Race, Gender, and Sexuality in the Colonial Contest* (New York: Routledge, 1995), proposes a "porno-tropic tradition" for this imperial narrative of possession.

10. Zamora, "Abreast of Columbus," 127–30; and Danièle Lavallée, *The First South Americans: The Peopling of a Continent from the Earliest Evidence to High Culture,* trans. Paul G. Bahn (Salt Lake City: University of Utah Press, 2000), 1.

11. Zamora, "Abreast of Columbus," 130.

12. *The Travels of Sir John Mandeville* (London: Macmillan, 1900), 127–28; *Log of Christopher Columbus,* 25; and Campbell, *Witness,* 10, 161. For a cautionary note on Mandeville's influence on Columbus, see Flint, *Imaginative Landscape,* 99–104.

13. An idiosyncratic tracing of the European search for Paradise on earth rooted in the author's postcolonial "predicament" is Henri Baudet, *Paradise on Earth: Some Thoughts on European Images of Non-European Man,* trans. Elizabeth Wentholt (New Haven, Conn.: Yale University Press, 1965).

14. Lavallée, *First South Americans,* 2; and Jan Rogozinski, *A Brief History of the Caribbean: From the Arawak and the Carib to the Present* (New York: Facts On File, 1999), 26.

15. Besides the image of Asia, influential in the conduct of expansion were the particulars of the Spanish *reconquista* of the Iberian Peninsula, the roles of the church and crown, and European aspirations of the period more generally. J. H. Elliott, "The Spanish Conquest and Settlement of America," in *The Cambridge History of Latin America,* vol. 1, *Colonial Latin America,* ed. Leslie Bethell (Cambridge: Cambridge University Press, 1984), 149–62.

16. Lavallée, *First South Americans,* 6. On monstrous men and beasts, Amazons, and cannibals in medieval European and Columbus's accounts, see Flint, *Imaginative Landscape,* 16–17, 138–45; and Zamora, "Abreast of Columbus," 139–41.

17. Rogozinski, *Brief History,* 26; and Gary B. Nash and Julie Roy Jeffrey, *The American People: Creating a Nation and a Society,* 4th ed. (New York: Longman, 1998), 20–21.

18. Lavallée, *First South Americans,* 4, 6.

19. The pursuit of gold and slaves and the establishment of tropical plantations employing enslaved labor in Africa and its Atlantic islands, the Canaries, Madeira, and the Azores, set a pattern for the American experience. Elliott, "Spanish Conquest," 154–55.

20. For a brief discussion of the legal and theological grounds for slavery, see Elliott, "Spanish Conquest," 163.

21. John Hemming, *Red Gold: The Conquest of the Brazilian Indians* (Cambridge, Mass.: Harvard University Press, 1978), 45–46, 69. See also Lavallée, *First South Americans,* 6–9.

22. Benjamin Schmidt, *Innocence Abroad: The Dutch Imagination and the New World, 1570–1670* (Cambridge: Cambridge University Press, 2001), xvii, xviii, xix–xxiii.

23. C. R. Boxer, *The Dutch in Brazil, 1624–1654* (Oxford: Clarendon Press, 1957), 1–66.

24. Boxer, *Dutch in Brazil,* 69, 70, 112.

25. Boxer, *Dutch in Brazil,* 112–13, 152–54.

26. León Krempel, ed., *Frans Post (1612–1680): Painter of Paradise Lost* (Petersberg, Germany: Michael Imhof Verlag, 2006), 45.

27. Schmidt, *Innocence Abroad,* 311–13, 316–20. See also Erik Larsen, *Frans Post: Interprète du Brésil* (Amsterdam: Colibris, 1962); Joaquim De Sousa-Leão, *Frans Post, 1612–1680* (Amsterdam: A. L. Van Gendt, 1973); Pedro Corrêa do Lago and Blaise Ducos, eds., *Frans Post: Le Brésil à la cour de Louis XIV* (Milan: 5 Continents Edition, 2005); and Krempel, *Frans Post,* 45.

28. Krempel, *Frans Post,* 46–50.

29. Elliott, "Spanish Conquest," 163; and Lavallée, *First South Americans,* 8.

30. Immanuel Kant, *Physical Geography,* as excerpted in *Race and the Enlightenment: A Reader,* ed. Emmanuel Chukwudi Eze (Cambridge, Mass.: Blackwell, 1997), 63, 64; and G. W. F. Hegel, *Lectures on the Philosophy of History,* trans. J. Sibree (London: George Bell and Sons, 1881), 83, 84.

31. Humboldt's five-volume *Examen critique de l'histoire de la géographie du Nouveau Continent et des progrès de l'astronomie nautique aux quinzième et seizième siècles* (Paris, 1836–39) evaluated the discoveries of Columbus and Vespucci in the light of modern geography.

32. David Arnold, "'Illusory Riches': Representations of the Tropical World, 1840–1950," *Singapore Journal of Tropical Geography* 21:1 (2000): 8. For a perspective on Humboldt's project, see Michael Dettelbach, "Global Physics and Aesthetic Empire: Humboldt's

Physical Portrait of the Tropics," in *Visions of Empire: Voyages, Botany, and Representations of Nature,* ed. David Philip Miller and Peter Hanns Reill (Cambridge: Cambridge University Press, 1996), 258–92.

33. Alexander von Humboldt, *Personal Narrative of Travels to the Equinoctial Regions of America, during the Years 1799–1804,* vol. 1, trans. and ed. Thomasina Ross (London: George Bell and Sons, 1889), x, xxi.

34. Humboldt, *Personal Narrative,* vol. 1, 1–2, 133–34, 135.

35. Humboldt, *Personal Narrative,* vol. 1, 136–41; see also p. 31 for the influence of climate on bodily forms.

36. Humboldt, *Personal Narrative,* vol. 1, 141, 142. On acclimatization, see David N. Livingstone, "Human Acclimatization: Perspectives on a Contested Field of Inquiry in Science, Medicine and Geography," *History of Science* 25:4 (1987): 364–66; and Warwick Anderson, "Climates of Opinion: Acclimatization in Nineteenth-Century France and England," *Victorian Studies* 35:2 (1992): 135–57.

37. Humboldt, *Personal Narrative,* vol. 1, 147.

38. Humboldt, *Personal Narrative,* vol. 1, 143–44, 194, 206, 304; and vol. 3, 95–96.

39. For a more ancient origin of the noble savage in the European idea of "good oriental" and its much later reappearance in the natives of America, see Baudet, *Paradise on Earth,* 21–22, 26–28.

40. Humboldt, *Personal Narrative,* vol. 1, 206–7. On the feminization of the land and men's desires for its subjugation, possession, and control, albeit of different times and places, see Annette Kolodny, *The Lay of the Land: Metaphor as Experience and History in American Life and Letters* (Chapel Hill: University of North Carolina Press, 1975). On sex and conquest, see Ronald Hyam, *Empire and Sexuality: The British Experience* (Manchester, England: Manchester University Press, 1990); and Richard C. Trexler, *Sex and Conquest: Gendered Violence, Political Order, and the European Conquest of the Americas* (Ithaca, N.Y.: Cornell University Press, 1995).

41. Humboldt, *Personal Narrative,* vol. 3, 216–17.

42. D. A. Brading, *The First America: The Spanish Monarchy, Creole Patriots, and the Liberal State, 1492–1867* (Cambridge: Cambridge University Press, 1991), 534.

43. Humboldt, *Personal Narrative,* vol. 3, 265.

44. Humboldt, *Personal Narrative,* vol. 3, 404, 407.

45. As quoted in Gertrude Himmelfarb, *Darwin and the Darwinian Revolution* (New York: W. W. Norton, 1959), 46, 47.

46. Elizabeth S. Eustis, *European Pleasure Gardens: Rare Books and Prints of Historic Landscape Design from the Elizabeth K. Reilley Collection* (New York: New York Botanical Garden, 2003), 10.

47. John Prest, *The Garden of Eden: The Botanic Garden and the Re-Creation of Paradise* (New Haven, Conn.: Yale University Press, 1981), 6, 21, 24, 27–37; and Martin Kemp, " 'Implanted in our Natures': Humans, Plants, and the Stories of Art," in Miller and Reill, *Visions of Empire,* 207–9. Ancient Greek ideas of paradise on earth may have derived from Indian Sanskrit writings, and European botanical gardens from the botanical gardens of late medieval Egypt, Arabia, and India.

48. Eustis, *European Pleasure Gardens,* 9–11; and Prest, *Garden of Eden,* 89.

49. For an illuminating instance of cultured ignorance, or "agnotology," amidst drives for systematics in an age of empire, see Londa Schiebinger, *Plants and Empire: Colonial Bioprospecting in the Atlantic World* (Cambridge, Mass.: Harvard University Press, 2004).

50. Livingstone, "Human Acclimatization," 364–66; and Anderson, "Climates of Opinion," 135–57.

51. For a survey of Banksian works, see Harold B. Carter, *Sir Joseph Banks, 1743–1820* (London: British Museum, 1988).

52. Schiebinger, *Plants and Empire,* 5, 6.

53. Lynne Withey, *Voyages of Discovery: Captain Cook and the Exploration of the Pacific* (Berkeley: University of California Press, 1987), 84–85.

54. For an account of this affair, see Withey, *Voyages of Discovery,* 169–96; and David Philip Miller, "Joseph Banks, Empire, and 'Centers of Calculation' in Late Hanoverian London," in Miller and Reill, *Visions of Empire,* 25–26.

55. Miller, "Joseph Banks," 27–31.

56. Alan Frost, "The Antipodean Exchange: European Horticulture and Imperial Designs," in Miller and Reill, *Visions of Empire,* 58–65. On Kew Gardens, see Lucile H. Brockway, *Science and Colonial Expansion: The Role of the British Royal Botanic Gardens* (New York: Academic Press, 1979).

57. Frost, "Antipodean Exchange," 74–75.

58. David Mackay, *In the Wake of Cook: Exploration, Science & Empire, 1780–1801* (New York: St. Martin's Press, 1985), 18.

59. David Mackay, "Agents of Empire: The Banksian Collectors and Evaluation of New Lands," in Miller and Reill, *Visions of Empire,* 39, 43.

60. Mackay, *In the Wake of Cook,* 123–43.

61. Schiebinger, *Plants and Empire,* 11–12.

62. See Mackay, *In the Wake of Cook,* 144–67.

63. Mackay, *In the Wake of Cook,* 12–13; and Mackay, "Agents of Empire," 54.

64. Mackay, "Agents of Empire," 54.

3. TROPICAL FRUIT

1. "An Industrial Empire Builder," *Outlook,* April 26, 1916, 990.

2. Charles David Kepner Jr. and Jay Henry Soothill, *The Banana Empire: A Case Study of Economic Imperialism* (New York: Vanguard Press, 1935), 336, 341.

3. Harry Elmer Barnes, "Introduction," in Kepner and Soothill, *Banana Empire,* 14. More than a "banana empire," Allan Punzalan Isaac has called that imperial discourse and possession the "American tropics" in his *American Tropics: Articulating Filipino America* (Minneapolis: University of Minnesota Press, 2006).

4. Barnes, "Introduction," 7–8.

5. See, e.g., Alan Kraut, *Silent Travelers: Germs, Genes and the "Immigrant Menace"* (New York: HarperCollins, 1994). On exchanges between America and Europe, see Alfred W. Crosby Jr., *The Columbian Exchange: Biological and Cultural Consequences of 1492* (Westport, Conn.: Greenwood Press, 1972); and Robert S. Desowitz, *Who Gave Pinta to the Santa Maria? Torrid Diseases in a Temperate World* (New York: W. W. Norton, 1997).

6. John Duffy, *The Sanitarians: A History of American Public Health* (Urbana: University of Illinois Press, 1990), 15, 20, 24–25, 26.

7. J. H. Powell, *Bring Out Your Dead: The Great Plague of Yellow Fever in Philadelphia in 1793* (Philadelphia: University of Pennsylvania Press, 1949), 4.

8. Powell, *Bring Out Your Dead,* 13, 148.

9. As quoted in William A. Petri Jr., "America in the World: 100 Years of Tropical Medicine and Hygiene," Presidential Address, *American Journal of Tropical Medicine and Hygiene* 71:1 (2004): 2.

10. Jones and Allen wrote *A Narrative of the Proceedings of the Black People, during the Late Awful Calamity in Philadelphia, in the Year 1793: and a Refutation of Some Censures, Thrown upon Them in Some Late Publications* (Philadelphia: William W. Woodward, 1794). See Phillip Lapsansky, "'Abigail, a Negress'": The Role and the Legacy of African

Americans in the Yellow Fever Epidemic," in *A Melancholy Scene of Devastation: The Public Response to the 1793 Philadelphia Yellow Fever Epidemic,* ed. J. Worth Estes and Billy G. Smith (Canton, Mass.: Science History Publications, 1997), 61–78.

11. Powell, *Bring Out Your Dead,* 38–39, 41–44, 282–84. See also Estes and Smith, *Melancholy Scene.*

12. Duffy, *Sanitarians,* 38–50.

13. Petri, "America in the World," 3, 7.

14. Petri, "America in the World," 2–4, 8–14.

15. As quoted in François Delaporte, *The History of Yellow Fever: An Essay on the Birth of Tropical Medicine,* trans. Arthur Goldhammer (Cambridge, Mass.: MIT Press, 1991), 140.

16. Petri, "America in the World," 15. That claim of medical discovery and advance is an exertion of power of the United States (Cuba Commission) over Cuba (Carlos Finlay) coincident with the island's colonization. Delaporte, *History of Yellow Fever,* 137–43.

17. Donald S. Burke, "American Society of Tropical Medicine and Hygiene: Centennial Celebration Address," Philadelphia, December 3, 2003, 2.

18. The Cuban Carlos Finlay was a graduate of the Jefferson Medical College of Philadelphia. The earliest use of the term "tropical diseases" appears in English navy surgeon Benjamin Mosely's *A Treatise on Tropical Diseases; on Military Operations; and on the Climate of the West Indies* (London, 1787). For an account on the "tropicalization" of diseases, see Stepan, *Picturing Tropical Nature,* 149–79.

19. "Study of Tropical Diseases," *Baptist Missionary Magazine* 87:3 (1907): 101; and "The Harvard School of Tropical Medicine," *Outlook,* October 18, 1913, 343. See also Arthur Inkersley, "An American School of Tropical Medicine," *Overland Monthly and Out West Magazine* 57:4 (1911): 373–75.

20. "Diseases of Hot Climates," *Medical News* 83:2 (1903): 84–86; and Zachary Gussow, *Leprosy, Racism, and Public Health: Social Policy in Chronic Disease Control* (Boulder, Colo.: Westview Press, 1989), 111–12.

21. "Medical Science and the Tropics," *Bulletin of the American Geographical Society of New York* 45:1 (1913): 435–38. On the costs to occupying soldiers in the tropics and their consequences, see Philip D. Curtin, *Death by Migration: Europe's Encounter with the Tropical World in the Nineteenth Century* (Cambridge: Cambridge University Press, 1989).

22. See, e.g., Vicente Navarro, ed., *Imperialism, Health and Medicine* (London: Pluto Press,

1982); and Roy MacLeod and Milton Lewis, eds., *Disease, Medicine, and Empire: Perspectives on Western Medicine and the Experience of European Expansion* (London: Routledge, 1988).

23. On human, plant, and animal acclimatization in nineteenth-century European thought and practice, see Livingstone, "Human Acclimatization," 359–94; and Anderson, "Climates of Opinion," 135–57.

24. Chas. E. Woodruff, *The Effects of Tropical Light on White Men* (New York: Rebman, 1905), v, 1, 4–7, 143–45, 190–203, 353. See also James Johnson, *The Influence of Tropical Climates on European Constitutions: Being a Treatise on the Principal Diseases Incidental to Europeans in the East and West Indies, Mediterranean, and Coast of Africa* (New York: Duyckinck, Long, Collins and Company, 1826).

25. Mary Endicott Chamberlain, "An Obligation of Empire," *North American Review* 170:71 (1900): 494–95.

26. As quoted in Nash and Jeffrey, *American People,* 703.

27. See, e.g., Thomas J. Osborne, *"Empire Can Wait": American Opposition to Hawaiian Annexation, 1893–1898* (Kent, Ohio: Kent State University Press, 1981); Richard E. Welch Jr., *Response to Imperialism: The United States and the Philippine-American War, 1899–1902* (Chapel Hill: University of North Carolina Press, 1979); and E. Berkeley Thompkins, *Anti-Imperialism in the United States: The Great Debate, 1890–1920* (Philadelphia: University of Pennsylvania Press, 1970).

28. See, e.g., Gail Bederman, *Manliness and Civilization: A Cultural History of Gender and Race in the United States, 1880–1917* (Chicago: University of Chicago Press, 1995); and Kristin L. Hoganson, *Fighting for American Manhood: How Gender Politics Provoked the Spanish-American and Philippine-American Wars* (New Haven, Conn.: Yale University Press, 1998).

29. For a discussion of overlapping empires, British and U.S., and Anglo-Saxonism, see Paul A. Kramer, "Empires, Exceptions, and Anglo-Saxons: Race and Rule between the British and U.S. Empires, 1880–1910," in *The American Colonial State in the Philippines: Global Perspectives,* ed. Julian Go and Anne L. Foster (Durham, N.C.: Duke University Press, 2003), 43–91.

30. *Rudyard Kipling's Verse, 1885–1926* (Garden City, N.Y.: Doubleday, Page and Company, 1927), 373–74.

31. As quoted in Elizabeth Mary Holt, *Colonizing Filipinas: Nineteenth-Century Representations of the Philippines in Western Historiography* (Manila: Ateneo de Manila University Press, 2002), 60.

32. Nash and Jeffrey, *American People,* 704.

33. Mark Twain, "To the Person Sitting in Darkness," February 1901, as reprinted in *Vestiges of War: The Philippine-American War and the Aftermath of an Imperial Dream, 1899–1999,* ed. Angel Velasco Shaw and Luis H. Francia (New York: New York University Press, 2002), 64.

34. As quoted in Holt, *Colonizing Filipinas,* 60–61, 74–80.

35. As quoted in Matthew Frye Jacobson, *Barbarian Virtues: The United States Encounters Foreign Peoples at Home and Abroad, 1876–1917* (New York: Hill and Wang, 2000), 229, 232.

36. Jacobson, *Barbarian Virtues,* 56. On this relationship between export markets and labor markets, empire and immigration, see pp. 11–97.

37. See, e.g., Marcus Lee Hansen, *The Atlantic Migration, 1607–1860: A History of the Continuing Settlement of the United States,* ed. Arthur M. Schlesinger (Cambridge, Mass.: Harvard University Press, 1940); and Oscar Handlin, *The Uprooted: The Epic Story of the Great Migrations That Made the American People* (New York: Grosset and Dunlop, 1951).

38. Madison Grant, "Introduction," in *The Rising Tide of Color against White World-Supremacy,* by Lothrop Stoddard (New York: Charles Scribner's Sons, 1920), xi.

39. Prescott F. Hall, "Immigration Restriction and World Eugenics," *Journal of Heredity* 10:3 (1919): 126, and as quoted in Stoddard, *Rising Tide,* 259–60. Hall and other Harvard graduates formed the Immigration Restriction League in 1894, with Hall serving as its secretary.

40. Stoddard, *Rising Tide,* 261, 262, 263. In an earlier book, Stoddard lamented the "loss" of Haiti in "the world-wide struggle between the primary races of mankind" or the "conflict of color," which was "the fundamental problem of the twentieth century." T. Lothrop Stoddard, *The French Revolution in San Domingo* (Boston: Houghton Mifflin, 1914), vii.

41. See, e.g., Hall, *Immigration and Its Effects,* which devotes part 4 to Chinese immigration.

42. Gussow, *Leprosy, Racism,* 19–22, 46, 85–107, 111–29. For Hawaiian reactions to leprosy and banishment, see Gussow, *Leprosy, Racism,* 96–101; and *The True Story of Kaluaikoolau as Told by His Wife, Piilani,* trans. Frances N. Frazier (Lihuʻi: Kauaʻi Historical Society, 2001).

43. Nayan Shah, *Contagious Divides: Epidemics and Race in San Francisco's Chinatown* (Berkeley: University of California Press, 2001), 1–2, 127–29. See also James C. Mohr, *Plague and Fire: Battling Black Death and the 1900 Burning of Honolulu's Chinatown*

(New York: Oxford University Press, 2005); and Tin-Yuke Char, ed., *The Sandalwood Mountains: Readings and Stories of the Early Chinese in Hawaii* (Honolulu: University of Hawai'i Press, 1975), 101–10.

44. Gussow, *Leprosy, Racism,* 129.

45. For a review of this literature, see Gary Y. Okihiro, *The Columbia Guide to Asian American History* (New York: Columbia University Press, 2001), 194–203. For a perspective on the "yellow" and allied "perils," see Gary Y. Okihiro, *Margins and Mainstreams: Asians in American History and Culture* (Seattle: University of Washington Press, 1994), 118–47.

46. Pierton W. Dooner, *Last Days of the Republic* (San Francisco: Alta California, 1879), 3, 15, 17, 96–97, 145–63, 209, 248, 257.

47. See, e.g., Thurston Clarke, *Pearl Harbor Ghosts: A Journey to Hawaii Then and Now* (New York: William Morrow, 1991); and Michael Crichton, *Rising Sun* (New York: Ballantine Books, 1992). Recent versions of future contests with China and Japan include Richard Bernstein and Ross H. Munro, *The Coming Conflict with China* (New York: Knopf, 1997); and George Friedman and Meredith Lebard, *The Coming War with Japan* (New York: St. Martin's Press, 1991). A more comprehensive explication of the dangers posed by Asians is Samuel P. Huntington, *The Clash of Civilizations and the Remaking of World Order* (New York: Simon and Schuster, 1996).

48. Clare Boothe, "Ever Hear of Homer Lea?" *Saturday Evening Post,* March 7, 1942, 12–13, 69–72, and March 14, 1942, 27, 38–40, 42. Clare Boothe went on to become Clare Boothe Luce, congresswoman and wife of the publisher of *Time, Life,* and *Fortune.* See also Wilbur Burton, "Prophet of Pearl Harbor," *Saturday Review of Literature* 25:14 (1942): 8. An earlier forecast of war with Japan was published by Richmond Pearson Hobson in the Sunday supplements of the *San Francisco Examiner,* November 3 and 10, 1907, and reprinted in *Cosmopolitan Magazine* 54:6 (1908): 584–93, and 55:1 (1908): 38–47.

49. Homer Lea, *The Valor of Ignorance* (New York: Harper and Brothers, 1909), 10–11, 23, 24.

50. Lea, *Valor of Ignorance,* 157–59, 170, 183, 307.

51. For a synopsis of this unity, see Okihiro, *Margins and Mainstreams,* 31–63. Substantial studies include Reginald Kearney, *African American Views of the Japanese: Solidarity or Sedition?* (Albany: State University of New York Press, 1998); Marc Gallicchio, *The African American Encounter with Japan and China: Black Internationalism in Asia, 1895–1945* (Chapel Hill: University of North Carolina Press, 2000); Vijay Prashad, *Everybody Was Kung Fu Fighting: Afro-Asian Connections and the Myth of Cultural*

Purity (Boston: Beacon Press, 2001); and Najia Aarim-Heriot, *Chinese Immigrants, African Americans, and Racial Anxiety in the United States, 1848–82* (Urbana: University of Illinois Press, 2003).

52. As quoted in Gerald Horne, *Race War! White Supremacy and the Japanese Attack on the British Empire* (New York: New York University Press, 2004), 45.

53. See, e.g., John W. Dower, *War without Mercy: Race and Power in the Pacific War* (New York: Pantheon, 1986); and for a more nuanced account of Japan's nationalism and expansionism, see Stefan Tanaka, *Japan's Orient: Rendering Pasts into History* (Berkeley: University of California Press, 1993); Horne, *Race War;* and T. Fujitani, Geoffrey M. White, and Lisa Yoneyama, eds., *Perilous Memories: The Asia-Pacific War(s)* (Durham, N.C.: Duke University Press, 2001).

54. For a broader canvas of world history, compare Immanuel Wallerstein, *The Modern World-System,* vol. 3, *The Second Era of Great Expansion of the Capitalist World-Economy, 1730–1840s* (New York: Academic Press, 1989), with Andre Gunder Frank, *ReOrient: Global Economy in the Asian Age* (Berkeley: University of California Press, 1998).

55. Dower, *War without Mercy,* 147–80; and Gallicchio, *African American Encounter,* 6–29.

56. As quoted in Gary Y. Okihiro, *Cane Fires: The Anti-Japanese Movement in Hawaii, 1865–1945* (Philadelphia: Temple University Press, 1991), 97.

57. Okihiro, *Cane Fires,* 117.

58. Okihiro, *Cane Fires,* details the execution of those plans, in contrast to Japan's war plans, as described in John J. Stephan, *Hawaii under the Rising Sun: Japan's Plans for Conquest after Pearl Harbor* (Honolulu: University of Hawai'i Press, 1984), which failed to materialize.

59. As quoted in Edward D. Beechert, *Working in Hawaii: A Labor History* (Honolulu: University of Hawai'i Press, 1985), 79.

60. For an account of the demographic debate, see Okihiro, *Columbia Guide,* 45–55.

61. Gavan Daws, *Shoal of Time: A History of the Hawaiian Islands* (Honolulu: University of Hawai'i Press, 1986), 211.

62. Quoted in David Northrup, *Indentured Labor in the Age of Imperialism, 1834–1922* (Cambridge: Cambridge University Press, 1995), 16. See also Alan H. Adamson, *Sugar without Slaves: The Political Economy of British Guiana, 1838–1904* (New Haven, Conn.: Yale University Press, 1972); Hugh Tinker, *A New System of Slavery: The Export of Indian Labour Overseas, 1830–1920* (London: Oxford University Press, 1974); Walton

Look Lai, *Indentured Labor, Caribbean Sugar: Chinese and Indian Migrants to the British West Indies, 1838–1918* (Baltimore, Md.: Johns Hopkins University Press, 1993); Moon-Ho Jung, *Coolies and Cane: Race, Labor, and Sugar in the Age of Emancipation* (Baltimore, Md.: Johns Hopkins University Press, 2006); and Lisa Yun, *The Coolie Speaks: Chinese Indentured Laborers and African Slaves in Cuba* (Philadelphia: Temple University Press, 2008).

63. As quoted in Clarence E. Glick, *Sojourners and Settlers: Chinese Migrants in Hawaii* (Honolulu: Hawaii Chinese History Center, 1980), 24. Contract terms varied from five to three years.

64. Quoted in Okihiro, *Cane Fires*, 19, 20.

65. On plantation workers and resistance, see Brij V. Lal, Doug Munro, and Edward D. Beechert, eds., *Plantation Workers: Resistance and Accommodation* (Honolulu: University of Hawai'i Press, 1993).

4. PINEAPPLE DIASPORA

1. J. L. Collins, "Pineapples in Ancient America," *Scientific Monthly* 67:5 (1948): 372.

2. See, e.g., Crosby, *Columbian Exchange;* and Brockway, *Science and Colonial Expansion.*

3. Berthold Laufer, "The American Plant Migration," *Scientific Monthly* 28:3 (1929): 241.

4. Barbara Pickersgill and Charles B. Heiser Jr., "Origins and Distribution of Plants Domesticated in the New World Tropics," in *Origins of Agriculture,* ed. Charles A. Reed (The Hague: Mouton, 1977), 803–35. There are competing claims, including ancient Assyria, Egypt, and Rome. See an assessment of these in J. L. Collins, *The Pineapple: Botany, Cultivation, and Utilization* (London: Leonard Hill, 1960), 18–20.

5. Kenneth F. Baker and J. L. Collins, "Notes on the Distribution and Ecology of Ananas and Pseudananas in South America," *American Journal of Botany* 26:9 (1939): 697–702.

6. Collins, *Pineapple,* 6, 32–35.

7. Collins, "Pineapples in Ancient America," 376; and Collins, *Pineapple,* 4–9, 32–35.

8. John Hemming, "The Indians of Brazil in 1500," in *The Cambridge History of Latin America,* vol. 1, *Colonial Latin America,* ed. Leslie Bethell (Cambridge: Cambridge University Press, 1984), 119; and William Balée, *Footprints of the Forest: Ka'apor Ethnobotany: The Historical Ecology of Plant Utilization by an Amazonian People* (New York: Columbia University Press, 1994), 138. For an account of early humans in this area, see Lavallée, *First South Americans,* 105–13, 149–54, 160–63.

9. David J. Wilson, *Indigenous South Americans of the Past and Present: An Ecological Perspective* (Boulder, Colo.: Westview Press, 1999), 82.

10. Whether they retained and passed on horticultural knowledge is a matter of debate. See, e.g., Balée, *Footprints of the Forest*, 138–41.

11. Wilson, *Indigenous South Americans*, 78–82. For a contemporary but related, detailed study of plants and the Ka'apor (a Tupí-Guaraní people), see Balée, *Footprints*.

12. Lavallée, *First South Americans*, 105. See also Wilson, *Indigenous South Americans*, 148–51.

13. Wilson, *Indigenous South Americans*, 1.

14. Balée, *Footprints of the Forest*, 49, 50, 64, 121–23, 179–81.

15. For a survey of Tupí society from European sources, see Hemming, "Indians of Brazil," 124–35.

16. John Hemming, *Red Gold,* 51–55; and Hemming, "Indians of Brazil," 120.

17. Hemming, "Indians of Brazil," 122, 125, 131–32, 134–35; Balée, *Footprints of the Forest,* 141–42; William Balée, "The Ecology of Ancient Tupi Warfare," in *Warfare, Culture, and Environment,* ed. R. Brian Ferguson (Orlando, Fla.: Academic Press, 1984), 241–65; and Jorge Hidalgo, "The Indians of Southern South America in the Middle of the Sixteenth Century," in Bethell, *Cambridge History,* 110.

18. Hemming, *Red Gold,* 49–50.

19. Mary W. Helms, "The Indians of the Caribbean and Circum-Caribbean at the End of the Fifteenth Century," in Bethell, *Cambridge History,* 37–44, 54–57; and Nelly Arvelo-Jiménez and Horacio Biord, "The Impact of Conquest on Contemporary Indigenous Peoples of the Guiana Shield: The System of Orinoco Regional Interdependence," in *Amazonian Indians from Prehistory to the Present: Anthropological Perspectives,* ed. Anna Roosevelt (Tucson: University of Arizona Press, 1994), 55–78.

20. Louis Allaire, "The Lesser Antilles before Columbus," in *The Indigenous Peoples of the Caribbean,* ed. Samuel M. Wilson (Gainesville: University Press of Florida, 1997), 20–28.

21. David R. Watters, "Maritime Trade in the Prehistoric Eastern Caribbean," in Wilson, *Indigenous Peoples,* 88.

22. James B. Petersen, "Taino, Island Carib, and Prehistoric Amerindian Economies in the West Indies: Adaptations to Island Environments," in Wilson, *Indigenous Peoples,* 118–27, 128; and Helms, "Indians of the Caribbean," 52–55.

23. Collins, "Pineapples in Ancient America," 372; and Collins, *Pineapple*, 4. On this voyage, Columbus introduced sugarcane to America. Philip D. Curtin, *The Rise and Fall of the Plantation Complex* (Cambridge: Cambridge University Press, 1990), 25.

24. Collins, *Pineapple*, 4–5; Collins, "Pineapples in Ancient America," 372–77; and Laufer, "American Plant Migration," 246–48. For an account of the pineapple's appearance in Europe, especially England, see Fran Beauman, *The Pineapple: King of Fruits* (London: Chatto and Windus, 2005).

25. Laufer, "American Plant Migration," 246, 247.

26. As quoted in Hemming, *Red Gold*, 223. For more on Raleigh's accounts, see Louis Montrose, "The Work of Gender in the Discourse of Discovery," in *New World Encounters*, ed. Stephen Greenblatt (Berkeley: University of California Press, 1993), 177–217; and Mary C. Fuller, "Raleigh's Fugitive Gold: Reference and Deferral in *The Discoverie of Guiana*," in Greenblatt, *New World Encounters*, 218–40.

27. As translated and quoted in Collins, *Pineapple*, 9–14.

28. Collins, *Pineapple*, 21.

29. Beauman claims the artist was possibly John Michael Wright. *Pineapple*, 49.

30. Laufer, "American Plant Migration," 248–49, 250–51; Collins, *Pineapple*, 18, 21–22; and J. L. Collins, "Notes on the Origin, History, and Genetic Nature of the Cayenne Pineapple," *Pacific Science* 5:1 (1951): 3.

31. As suggested by Beauman, *Pineapple*, 48.

32. The counterparts to European gardens were colonial gardens such as established by the Dutch at the Cape of Good Hope in 1694, the French on Mauritius in 1735, and the English in Jamaica, St. Vincent, Calcutta, and Penang in the early eighteenth century. Anderson, "Climates of Opinion," 137. See also Brockway, *Science and Colonial Expansion;* and Mackay, "Agents of Empire."

33. Apparently a Dutch physician, Bernadus Paladanus, tried to cultivate pineapples in his Enkhuizen garden in 1592 but failed because of the cold. Beauman, *Pineapple*, 58.

34. Collins, *Pineapple*, 22, 24–25; and Laufer, "American Plant Migration," 249.

35. Laufer, "American Plant Migration," 249.

36. Collins, *Pineapple*, 25, 68–69.

37. As translated in Collins, *Pineapple*, 10, 11, 12.

38. Louise Conway Belden, *The Festive Tradition: Table Decoration and Desserts in America, 1650–1900* (New York: W. W. Norton, 1983), 222.

39. Collins, *Pineapple,* 26.

40. This remarkable account is taken from Collins, "Notes on the Origin," 3–17; and Collins, *Pineapple,* 69–77. For the more general process of empire and plant transfers, see Brockway, *Science and Colonial Expansion.*

41. As quoted and translated in Collins, "Notes on the Origin," 6.

42. From an advertisement in the *Gardeners' Chronicle* (England), March 6, 1841.

43. Collins, "Notes on the Origin," 3, 4.

44. For a different account of the introduction of the Cayenne to Hawai'i, see Jan K. TenBruggencate, *Hawai'i's Pineapple Century: A History of the Crowned Fruit in the Hawaiian Islands* (Honolulu: Mutual, 2004), 5–6.

45. The papaya followed a similar migration route. By 1911, it had been taken by Europeans from the Caribbean to Hawai'i, where it was grown commercially; by 1936, Hawaiian papayas were sold in Japan and the West Coast. During the 1980s, Hawaiian papaya seeds and commercial farming techniques returned to the Caribbean, where papayas were grown and transported to Europe. Ian Cooke et al., "Follow the Thing: Papaya," *Antipode* 36:4 (2004): 652.

5. HAWAIIAN MISSION

The section titles "Native Land" and "Foreign Desires" come from Lilikala Kame'elei-hiwa, *Native Land and Foreign Desires* (Honolulu: Bishop Museum Press, 1992).

1. Davianna Pomaika'i McGregor, "The Cultural and Political History of Hawaiian Native People," in *Our History Our Way: An Ethnic Studies Anthology,* ed. Gregory Yee Mark, Davianna Pomaika'i McGregor, and Linda A. Revilla (Dubuque, Iowa: Kendall/Hunt, 1996), 335–36.

2. Patrick Vinton Kirch, *Feathered Gods and Fishhooks: An Introduction to Hawaiian Archaeology and Prehistory* (Honolulu: University of Hawai'i Press, 1985), 2; and E. S. Craighill Handy and Elizabeth Green Handy, *Native Planters in Old Hawaii: Their Life, Lore and Environment,* Bulletin 233 (Honolulu: Bishop Museum Press, 1972), 48–49, 73. On the origin and spread of the sweet potato, see Kirch, *Feathered Gods,* 65.

3. As quoted in Daws, *Shoal of Time,* 106.

4. For a discussion of the debate on Hawaiian demography, see Okihiro, *Columbia Guide* 45–55.

5. Ralph S. Kuykendall, *The Hawaiian Kingdom,* vol. 1, *Foundation and Transformation,*

1778–1854 (Honolulu: University of Hawai'i Press, 1938), 29; and Noel J. Kent, *Hawaii: Islands under the Influence* (New York: Monthly Review Press, 1983), 19.

6. Beechert, *Working in Hawaii*, 10–14; and Samuel Kamakau as quoted in Kuykendall, *Hawaiian Kingdom*, vol. 1, 88–89.

7. Robert C. Schmitt, *Demographic Statistics of Hawaii: 1778–1965* (Honolulu: University of Hawai'i Press, 1968), 39. On the variable enlistments of Hawaiian men on foreign ships, see Kuykendall, *Hawaiian Kingdom*, vol. 1, 312–13.

8. Kent, *Hawaii*, 18.

9. Kuykendall, *Hawaiian Kingdom*, vol. 1, 88, 89.

10. Kuykendall, *Hawaiian Kingdom*, vol. 1, 91–92.

11. Char, *Sandalwood Mountains*, 37.

12. Kuykendall, *Hawaiian Kingdom*, vol. 1, 93–95, 302, 304, 305–13. In addition to being agent in 1817, Hunnewell was first officer on board the *Thaddeus*, which conveyed the first American missionaries and their Hawaiian converts from Boston to the Islands in 1819–20. Sidney and Marjorie Barstow Greenbie, *Gold of Ophir: The China Trade in the Making of America* (New York: Wilson-Erickson, 1937), 122.

13. Warren L. Cook, *Flood Tide of Empire: Spain and the Pacific Northwest, 1543–1819* (New Haven, Conn.: Yale University Press, 1973), 107; and Arrell Morgan Gibson, *Yankees in Paradise: The Pacific Basin Frontier* (Albuquerque: University of New Mexico Press, 1993), 70.

14. Gibson, *Yankees in Paradise*, 48–49, 51–52, 63–68.

15. Gibson, *Yankees in Paradise*, 93–98.

16. Gibson, *Yankees in Paradise*, 101.

17. As quoted in Kuykendall, *Hawaiian Kingdom*, vol. 1, 101.

18. As quoted in Theodore Morgan, *Hawaii, A Century of Economic Change, 1778–1876* (Cambridge, Mass.: Harvard University Press, 1948), 92.

19. Paul William Harris, *Nothing but Christ: Rufus Anderson and the Ideology of Protestant Foreign Missions* (New York: Oxford University Press, 1999), 62–63.

20. The generation of funds was also influential in the mission boarding schools and their stress on a vocational education. See Okihiro, *Island World*, chapter 4.

21. Harris, *Nothing but Christ*, 66.

22. See, e.g., Titus Coan, *Life in Hawaii: An Autobiographical Sketch of Mission Life and Labors, 1835–1882* (New York: Anson D. F. Randolph, 1882), 42–60.

23. McGregor, "Cultural and Political History," 348; Kameʻeleihiwa, *Native Land,* 173–88; and Jonathan Kay Kamakawiwoʻole Osorio, *Dismembering Lāhui: A History of the Hawaiian Nation to 1887* (Honolulu: University of Hawaiʻi Press, 2002), 24–43, 67–104, 198–99.

24. The *kaunaoʻa* is a parasitic member of the morning glory family.

25. Samuel M. Kamakau, *Ruling Chiefs of Hawaii* (Honolulu: Kamehameha Schools Press, 1992), 369–70. Kamehameha III saw foreigners in his government as a way of preserving Hawaiian sovereignty against the threats posed by European nations. Kamakau, *Ruling Chiefs,* 401–3; and Kameʻeleihiwa, *Native Land,* 188–92.

26. As quoted in Kamakau, *Ruling Chiefs,* 399–401.

27. For a history of Honolulu, see Edward D. Beechert, *Honolulu: Crossroads of the Pacific* (Columbia: University of South Carolina Press, 1991).

28. As quoted in McGregor, "Cultural and Political History," 349–50. See also Kameʻeleihiwa, *Native Land,* 193–98.

29. Kameʻeleihiwa, *Native Land,* 64.

30. Kuykendall, *Hawaiian Kingdom,* vol. 1, 269. See also Kameʻeleihiwa, *Native Land,* 51–64.

31. As quoted in Kuykendall, *Hawaiian Kingdom,* vol. 1, 276. On the king's dilemma regarding native rights and foreign desires, see Kameʻeleihiwa, *Native Land,* 210.

32. On the Great Mahele, see Kuykendall, *Hawaiian Kingdom,* vol. 1, 269–98; Jon J. Chinen, *The Great Mahele: Hawaii's Land Division of 1848* (Honolulu: University of Hawaiʻi Press, 1958); Kameʻeleihiwa, *Native Land,* 201–318; and Osorio, *Dismembering Lāhui,* 44–73.

33. Neil M. Levy, "Native Hawaiian Land Rights," *California Law Review* 63:4 (1975): 856; McGregor, "Cultural and Political History," 351; and Kameʻeleihiwa, *Native Land,* 295–98.

34. Kameʻeleihiwa, *Native Land,* 206; Morgan, *Hawaii,* 138–39; and McGregor, "Cultural and Political History," 352.

35. As quoted in Levy, "Native Hawaiian," 858.

36. Richard Armstrong to R. Anderson, December 21, 1839, Wailuku, Maui, Folder, Richard Armstrong, October 1832–October 1835, ABCFM-Hawaii Papers, Houghton Library (Harvard), 1820–1900, Hawaiian Mission Children's Society, Honolulu.

37. Levy, "Native Hawaiian," 858.

38. Levy, "Native Hawaiian," 859–61. For critical assessments of the Bishop estate, see George Cooper and Gavan Daws, *Land and Power in Hawaii: The Democratic Years* (Honolulu: Benchmark Books, 1985); and Samuel P. King and Randall W. Roth, *Broken Trust: Greed, Mismanagement, and Political Manipulation at America's Largest Charitable Trust* (Honolulu: University of Hawai'i Press, 2006).

39. As cited in Donald Cutter, "The Spanish in Hawaii: Gaytan to Marin," *Hawaiian Journal of History* 14 (1980): 22–23.

40. Kent, *Hawaii,* 36–38; and Lawrence H. Fuchs, *Hawaii Pono: A Social History* (New York: Harcourt, Brace and World, 1961), 22–24.

41. Kuykendall, *Hawaiian Kingdom,* vol. 1, 175.

42. It is important to note that local Kaua'i chiefs opposed the deal, fearing the loss of their hold over their people, but their appeal was rejected by the king. Beechert, *Working in Hawaii,* 22.

43. As quoted in Ronald Takaki, *Pau Hana: Plantation Life and Labor in Hawaii, 1835–1920* (Honolulu: University of Hawai'i Press, 1983), 5. On the machinations of Ladd and Company, see Kame'eleihiwa, *Native Land,* 179–80.

6. TROPICAL PLANTATION

1. Kamakau, *Ruling Chiefs,* 425–26.

2. Morgan, *Hawaii,* 88–89.

3. Andre Gunder Frank, *ReOrient: Global Economy in the Asian Age* (Berkeley: University of California Press, 1998). See also Janet L. Abu-Lughod, *Before European Hegemony: The World System A.D. 1250–1350* (New York: Oxford University Press, 1989).

4. Lach, *Asia in the Making of Europe,* vol. 1, 15.

5. Curtin, *Rise and Fall,* 3–4. Memorably, the anthropologist Claude Lévi-Strauss wrote of the introduction of tropical foods into Europe, "The visual or olfactory surprises they [tropical foods] provided, since they were cheerfully warm to the eye or exquisitely hot on the tongue, added a new range of sense experience to a civilization which had never suspected its own insipidity." *Tristes tropiques,* trans. John and Doreen Weightman (New York: Penguin Books, 1973), 38.

6. J. H. Galloway, *The Sugar Cane Industry: An Historical Geography from Its Origins to 1914* (Cambridge: Cambridge University Press, 1989), 19–33.

7. Curtin, *Rise and Fall,* 5–8.

8. Curtin, *Rise and Fall,* 11–13; and Galloway, *Sugar Cane Industry,* 31–47.

9. Doug Munro, "Patterns of Resistance and Accommodation," in *Plantation Workers: Resistance and Accommodation,* ed. Brij V. Lal, Doug Munro, and Edward D. Beechert (Honolulu: University of Hawai'i Press, 1993), 1.

10. George L. Beckford, *Persistent Poverty: Underdevelopment in Plantation Economies of the Third World* (London: Zed Books, 1983), 12. See also the useful distinctions among encounters and plantations in Curtin, *Rise and Fall,* 11–16; the relationship between development (core) and underdevelopment (periphery) in Frank, *Capitalism and Underdevelopment;* and the broad context of world history in E.J. Hobsbawm, *The Age of Capital, 1848–1875* (London: Weidenfeld and Nicolson, 1975), 173–207; Fernand Braudel, *Civilization and Capitalism, 15th–18th Century,* vol. 3, *The Perspective of the World,* trans. Siân Reynolds (Berkeley: University of California Press, 1992); and Wallerstein, *Modern World-System,* vol. 3, 129–31, 137–38.

11. On the Atlantic sugar industry, see Curtin, *Rise and Fall;* Galloway, *Sugar Cane Industry;* and Sidney W. Mintz, *Sweetness and Power: The Place of Sugar in Modern History* (New York: Viking Press, 1985).

12. Morgan, *Hawaii,* 106–8, 150–57; and Kuykendall, *Hawaiian Kingdom,* vol. 1, 313–27.

13. As quoted in Kent, *Hawaii,* 36.

14. Some have claimed that the pineapple was introduced to Hawai'i by a Spaniard, Don Francisco de Paula Marin, who recorded growing it in 1813. See Blanche Kaualua L. Lee, *A History of the Events in the Life of Hawaii's Horticulturist: Don Francisco de Paula Marin* (Honolulu: Best Printing, 2002). Cf. Ross H. Gast, *Don Francisco de Paula Marin: A Biography* (Honolulu: University of Hawai'i Press, 1975), 19, 52, 209, 232.

15. Handy and Handy, *Native Planters,* 188; Kirch, *Feathered Gods,* 65; and TenBruggencate, *Hawai'i's Pineapple Century,* 1–2.

16. Withey, *Voyages of Discovery,* 214, 215, 294, 295, 412.

17. Kuykendall, *Hawaiian Kingdom,* vol. 1, 316; and Char, *Sandalwood Mountains,* 54–57.

18. Morgan, *Hawaii,* 179–80; and Ralph S. Kuykendall, *The Hawaiian Kingdom,* vol. 2, *Twenty Critical Years, 1854–1874* (Honolulu: University of Hawai'i Press, 1953), 141.

19. Ralph S. Kuykendall, *The Hawaiian Kingdom,* vol. 3, *The Kalakaua Dynasty, 1874–1893* (Honolulu: University of Hawai'i Press, 1967), 47, 83; and Daws, *Shoal of Time,* 208.

20. Kuykendall, *Hawaiian Kingdom,* vol. 1, 331–33; vol. 2, 145–49.

21. Kuykendall, *Hawaiian Kingdom,* vol. 2, 145–46.

22. Morgan, *Hawaii,* 186.

23. Daws, *Shoal of Time,* 312–13.

24. See, e.g., Noenoe K. Silva, *Aloha Betrayed: Native Hawaiian Resistance to American Colonialism* (Durham, N.C.: Duke University Press, 2004).

25. Osorio, *Dismembering Lāhui,* 168–69.

26. Hilary Conroy, *The Japanese Frontier in Hawaii, 1868–1898* (Berkeley: University of California Press, 1953), 57.

27. Kalākaua also proposed a Polynesian confederacy. Kuykendall, *Hawaiian Kingdom,* vol. 3, 305–39; and Osorio, *Dismembering Lāhui,* 229–35.

28. Conroy, *Japanese Frontier,* 51–53, 57; Daws, *Shoal of Time,* 217; and Kuykendall, *Hawaiian Kingdom,* vol. 3, 160.

29. The most comprehensive description of this period of Japanese migration to Hawai'i is Alan Takeo Moriyama, *Imingaisha: Japanese Emigration Companies and Hawaii, 1894–1908* (Honolulu: University of Hawai'i Press, 1985).

30. As quoted in Conroy, *Japanese Frontier,* 51–52; and Kuykendall, *Hawaiian Kingdom,* vol. 3, 228–31.

31. This crucial distinction is made by Osorio, *Dismembering Lāhui,* 146–47, 191–92.

32. McGregor, "Cultural and Political History," 359, 360; Osorio, *Dismembering Lāhui,* 145–92; and Silva, *Aloha Betrayed,* 87–122.

33. Osorio, *Dismembering Lāhui,* 180.

34. Osorio, *Dismembering Lāhui,* 193–249. Cf. Daws, *Shoal of Time,* 240–54.

35. Daws, *Shoal of Time,* 255–58; and Ernest Andrade Jr., *Unconquerable Rebel: Robert W. Wilcox and Hawaiian Politics, 1880–1903* (Niwot: University Press of Colorado, 1996), 49–67.

36. McGregor, "Cultural and Political History," 366.

37. Liliuokalani, *Hawaii's Story by Hawaii's Queen* (Rutland, Vt.: Charles E. Tuttle, 1964), 231.

38. McGregor, "Cultural and Political History," 367, 369–70. See also Lorrin A. Thurston, "The Sandwich Islands," *North American Review* 156:436 (1893): 270.

39. William Adam Russ Jr., *The Hawaiian Revolution (1893–94)* (Selinsgrove, Pa.: Susquehanna University Press, 1959), 77.

40. This account of the U.S.-abetted overthrow of the Hawaiian kingdom is taken from

McGregor, "Cultural and Political History," 365–73. See also Daws, *Shoal of Time,* 264–80; Silva, *Aloha Betrayed,* 123–35; and Russ, *Hawaiian Revolution,* 36–112.

41. As quoted in McGregor, "Cultural and Political History," 373–75.

42. McGregor, "Cultural and Political History," 375; and Helena G. Allen, *Sanford Ballard Dole: Hawaii's Only President, 1844–1926* (Glendale, Calif.: Arthur H. Clark, 1988), 207–8.

43. Andrade, *Unconquerable Rebel,* 149–66.

44. Two of those patriotic groups were the Hui Hawai'i Aloha 'Āina, comprising more than 7,500 native-born Hawaiian voters, and its women's branch of 11,000 members. Silva, *Aloha Betrayed,* 130–31, and 123–63 for an account of the grassroots organizing against U.S. annexation.

45. Liliuokalani, *Hawaii's Story,* 280–81, 289, 305, 323, 325; and Jonathan K. Kamakawiwo'ole Osorio, "A Hawaiian Nationalist Commentary on the Trial of the *Mo'iwahine,*" in *Trial of a Queen: 1895 Military Tribunal* (Honolulu: Judiciary History Center, 1996), 30. On Lili'uokalani's fight against U.S. annexation, see Silva, *Aloha Betrayed,* 164–203.

46. For a history of these Hawaiians, see Okihiro, *Island World,* chapters 3 and 5.

47. Liliuokalani, *Hawaii's Story,* 350–51. On cultural resistance to annexation, see Silva, *Aloha Betrayed,* 182–91; and Amy K. Stillman, "History Reinterpreted in Song: The Case of the Hawaiian Counterrevolution," *Hawaiian Journal of History* 23 (1989): 1–30.

48. Liliuokalani, *Hawaii's Story,* 354.

49. As quoted in Allen, *Sanford Ballard Dole,* 222.

50. Thurston, "Sandwich Islands," 271, 279, 280.

51. Daws, *Shoal of Time,* 293.

52. Richard Dole and Elizabeth Dole Porteus, *The Story of James Dole* ('Aiea, Hawai'i: Island Heritage, 1990), 21–22.

53. Campbell MacCulloch, "The Man Who Made Hawaii," *New McClure's* 62:3 (1929): 21.

7. HAWAIIAN PINE

1. As quoted in Dole and Porteus, *Story of James Dole,* 24.

2. Dole and Porteus, *Story of James Dole,* 25–27.

3. Glick, *Sojourners and Settlers,* 45–66. For the West, see Sucheng Chan, *This Bitter-Sweet*

Soil: The Chinese in California Agriculture, 1860–1910 (Berkeley: University of California Press, 1986).

4. Dole and Porteus, *Story of James Dole,* 29.

5. As cited in Gus M. Oehm, "By Nature Crowned: King of Fruits, Pineapples in Hawaii," Pineapple Research Institute, Honolulu, 1953, manuscript, Special Collections, Hamilton Library, University of Hawai'i, Mānoa, 112.

6. TenBruggencate, *Hawai'i's Pineapple Century,* 5. On early attempts to grow and market pineapple commercially, see E. C. Auchter, "People, Research, and Social Significance of the Pineapple Industry of Hawaii," Pineapple Research Institute, Honolulu, 1951, manuscript, Special Collections, Hamilton Library, University of Hawai'i, Mānoa; and Oehm, "By Nature Crowned."

7. TenBruggencate, *Hawai'i's Pineapple Century,* 6, 7. Englishman John Kidwell arrived in Hawai'i in 1882 and shortly thereafter began the first pineapple plantation in the Islands, on O'ahu. Oehm, "By Nature Crowned," 36–40. See also John Kidwell, "The Cultivation of Pineapples in Hawaii," *Hawaiian Forester and Agriculturist* 1:12 (1904): 334–45.

8. For another version, see Dole and Porteus, *Story of James Dole,* 35–46.

9. Dole and Porteus, *Story of James Dole,* 27, 38; and Oehm, "By Nature Crowned," 113.

10. TenBruggencate, *Hawai'i's Pineapple Century,* 27.

11. Dole and Porteus, *Story of James Dole,* 27.

12. Dole and Porteus, *Story of James Dole,* 38, 42.

13. Dole and Porteus, *Story of James Dole,* 42, 45, 50, 64, 65.

14. Dole and Porteus, *Story of James Dole,* 66, 67, 76.

15. Dole and Porteus, *Story of James Dole,* 68, 69, 70; and TenBruggencate, *Hawai'i's Pineapple Century,* 28. TenBruggencate states that Dole's interest in Haiku Fruit and Packing began in 1912 (p. 38).

16. For accounts of the privatization of Lana'i, see K. P. Emory, "Lanai Notes," Cabinet 1, Drawer 1, Folder 27, Lanai, Families—1930, 1950; and Marguerite K. Ashford, "Lanai: A Narrative History," Cabinet 1, Drawer 1, Folder 28, Lanai: A Narrative History by M. K. Ashford—1974, Dole Corporation Archives, Special Collections, Hamilton Library, University of Hawai'i, Mānoa. See also Lawrence Kainoahou Gay, *True Stories of the Island of Lanai* (Honolulu: Mission Press, 1965); and Elaine Kauwenaole Kaopuiki and Randolph Jordan Moore, *Lana'i, The Mystery Island* (Honolulu: Lopa, 1987).

17. Dole and Porteus, *Story of James Dole,* 71–73.

18. Gay, *True Stories,* 83. Cf. the gloss given by Hazel Carter Maxon, "A Deserted Island That Became a Pineapple Plantation," *Overland Monthly and Out West Magazine* 85:10 (1927): 296–98.

19. Kaopuiki and Moore, *Lana'i,* 32. See also Ashford, "Lanai," 90, 91.

20. Fuchs, *Hawaii Pono,* 249; and Caroline Manning, *The Employment of Women in the Pineapple Canneries of Hawaii,* Bulletin No. 82, Women's Bureau, U.S. Department of Labor (Washington, D.C.: Government Printing Office, 1930), 6.

21. For details on the seasonal nature of cannery work and differential wages by gender, see Manning, *Employment of Women,* 11–14, 20–25; *Labor Conditions in the Territory of Hawaii, 1929–1930,* Bulletin No. 534, Bureau of Labor Statistics, U.S. Department of Labor (Washington, D.C.: Government Printing Office, 1931), 77–79, 69–70; and Ethel Erickson, *Earnings and Hours in Hawaii Woman-Employing Industries,* Bulletin No. 177, Women's Bureau, U.S. Department of Labor (Washington, D.C.: Government Printing Office, 1940), 3–4, 6–14.

22. Beechert, *Working in Hawaii,* 181–83. For a personal account, see Chung Kun Ai, *My Seventy-Nine Years in Hawaii* (Causeway Bay, Hong Kong: Cosmorama Pictorial, 1960), 212–49.

23. Royal N. Chapman, *Cooperation in the Hawaiian Pineapple Business* (New York: Institute for Pacific Relations, 1933), 6.

24. On women in fieldwork, see Manning, *Employment of Women,* 4–5.

25. Edward Norbeck, *Pineapple Town: Hawaii* (Berkeley: University of California Press, 1959), 20–21, 22, 23.

26. Ida Kanekoa Milles, "Getting Somewheres," in *Hanahana: An Oral History Anthology of Hawaii's Working People,* ed. Michi Kodama-Nishimoto, Warren S. Nishimoto, and Cynthia A. Oshiro (Honolulu: Ethnic Studies Oral History Project, University of Hawai'i at Mānoa, 1984), 11–12.

27. As quoted in Dole and Porteus, *Story of James Dole,* 78, 79.

28. Manning, *Employment of Women,* 19; and Erickson, *Earnings and Hours,* 5.

29. Erickson, *Earnings and Hours,* 5.

30. Erickson, *Earnings and Hours,* 5, 13.

31. Fuchs, *Hawaii Pono,* 63–64; and Ashford, "Lanai," 88–89.

32. "Pineapples in Paradise," *Fortune* 11:5 (1930): 37.

33. On plantation paternalism in Hawai'i, see Beechert, *Working in Hawaii,* 177–95.

Similarly, industrial paternalism extended from factory to home, as when Henry Ford reportedly visited his workers' homes to ensure that they adhered to his values of domesticity. Stuart Ewen, *Captains of Consciousness* (New York: McGraw Hill, 1976), 133.

34. Moon-Kie Jung, *Reworking Race: The Making of Hawaii's Interracial Labor Movement* (New York: Columbia University Press, 2006), 47.

35. Beechert, *Working in Hawaii,* 182, 207.

36. Beechert, *Working in Hawaii,* 302–3; Sanford Zalburg, *A Spark Is Struck! Jack Hall and the ILWU in Hawaii* (Honolulu: University of Hawai'i Press, 1979), 171–86, 311–19; and TenBruggencate, *Hawai'i's Pineapple Century,* 149–50.

37. Dole and Porteus, *Story of James Dole,* 54.

38. "No you have never tasted pineapple," *Ladies' Home Journal,* January 1909, 42.

39. *Ladies' Home Journal,* June 1926, 170. See also *Ladies' Home Journal,* August 1926, 117; October 1926, 258; December 1926, 155; February 1927, 196; September 1972, 197; and November 1927, 170, for advertisements with personal testimonies of pineapple recipes.

40. *Hawaiian Pineapple as 100 Good Cooks Serve It* (San Francisco: Association of Hawaiian Pineapple Canners, 1928), 5. Other food companies published recipe books and sponsored contests for product and brand-name recognition. See, e.g., Carol Fisher, *The American Cookbook: A History* (Jefferson, N.C.: McFarland and Company, 2006), 126–51.

41. Dole and Porteus, *Story of James Dole,* 59. California's orange industry followed Hawaiian pineapple in generic and place-bound advertising.

42. *Ladies' Home Journal,* May 1931, 117.

43. *Ladies' Home Journal,* October 1, 1910, 82; November 1, 1910, 69; and January 1, 1911, 47.

44. *Ladies' Home Journal,* March 1916, 83; and July 1926, 95.

45. *Ladies' Home Journal,* January 1931, 115; and February 1931, 129.

46. Jennifer Saville, *Georgia O'Keeffe: Paintings of Hawai'i* (Honolulu: Honolulu Academy of Arts, 1990), 11, 12. Besides O'Keeffe, Dole through Ayer featured in its advertising works by contemporary artists Yasuo Kuniyoshi and Millard Sheets, and it invited Isamu Noguchi to Hawai'i for a marble bas-relief for the company's reception room. Over budget, the plan was abandoned by February 1941. Saville, *Georgia O'Keeffe,* 12; and correspondence from Amy Hau, administrative director, Noguchi Museum,

August 11, 2006. See also "Isamu Noguchi Sculptor, Here for Special Work," *Honolulu Star-Bulletin,* May 8, 1940; and "Noguchi Exhibit Opens," *Nippu Jiji,* May 14, 1940.

47. Neil Harris, "Designs on Demand: Art and the Modern Corporation," in *Art, Design, and the Modern Corporation* (Washington, D.C.: Smithsonian Institution Press, 1985), 8, 9. See also Andreas Huyssen, *After the Great Divide: Modernism, Mass Culture, Postmodernism* (Bloomington: Indiana University Press, 1986).

48. Saville, *Georgia O'Keeffe,* 13, 14.

49. Saville, *Georgia O'Keeffe,* 17.

50. "Pineapples in Paradise," *Fortune* 11:5 (1930): 33, 34, 36.

51. Henry A. White, *James D. Dole: Industrial Pioneer of the Pacific, Founder of Hawaii's Pineapple Industry* (New York: Newcomen Society in North America, 1957), 7, 8, 17, 20, 27, 28.

52. TenBruggencate, *Hawai'i's Pineapple,* 153–54, 155, 156, 157.

8. PINEAPPLE MODERN

1. See, e.g., "Dangers of Canned Food," *Medical News,* September 8, 1883, 270–71; "Effects of Heat and Cold on Canned Foods," *Scientific American,* May 27, 1893, 325; and Marcia Clarke, "Tinned Food for Tin Gods," *Forum,* February 1930, 94–96. Cf. Josephine Grenier, "New Uses for Canned Fruits," *Harper's* 43:3 (1909): 266–68; and "Canned Pineapple Recipes," *Christian Observer,* March 30, 1910, 18.

2. J. Alexis Shriver, *Pineapple-Canning Industry of the World,* Special Agents Series No. 91, U.S. Department of Commerce (Washington, D.C.: Government Printing Office, 1915), 7.

3. *The Story of the Canning Industry* (Washington, D.C.: National Canners Association, 1940), 3. See also Felipe Fernández-Armesto, *Food: A History* (London: Macmillan, 2001), 239, on Appert's invention and its military application.

4. Beauman, *Pineapple,* 212.

5. *Story of the Canning Industry,* 4; "Fruit Canning," *Massachusetts Ploughman and New England Journal of Agriculture,* January 15, 1870, 1; and Richard Hawkins, "The Baltimore Canning Industry and the Bahamian Pineapple Trade, c. 1865–1926," *Maryland Historian* 26:2 (1995): 3.

6. "Effects of Heat and Cold," *Scientific American,* May 27, 1893, 325.

7. "The American Can Company," *Fortune,* November 1930, 40–41.

8. Clarke, "Tinned Food," 94, 95; and "Romance in the Canning Industry," *Current Opinion,* November 1, 1924, 642.

9. Shriver, *Pineapple-Canning Industry,* 12, 17.

10. Roland Marchand, *Advertising the American Dream: Making Way for Modernity, 1920–1940* (Berkeley: University of California Press, 1985), 1–24.

11. For surveys of the advertising industry, see Joseph J. Selden, *The Golden Fleece: Selling the Good Life to Americans* (New York: Macmillan, 1963); Daniel Pope, *The Making of Modern Advertising* (New York: Basic Books, 1983); and Marchand, *Advertising.*

12. Marchand, *Advertising,* 66–69.

13. As quoted in Jennifer Scanlon, *Inarticulate Longings: The* Ladies' Home Journal, *Gender, and the Promises of Consumer Culture* (New York: Routledge, 1995), 10.

14. Also targeted were immigrant and working-class women who aspired to middle-class status. See Roger Miller, "*Selling Mrs. Consumer:* Advertising and the Creation of Suburban Socio-Spatial Relations, 1910–1930," *Antipode* 23:3 (1991): 263–306.

15. Mary Drake McFeely, *Can She Bake a Cherry Pie? American Women and the Kitchen in the Twentieth Century* (Amherst: University of Massachusetts Press, 2000), 34–50.

16. Christine Frederick, *Selling Mrs. Consumer* (New York: Business Bourse, 1929). See also Carl Naether, *Advertising to Women* (New York: Prentice Hall, 1928).

17. For a useful study on women and advertising, see Karen Elizabeth Altman, "Modernity, Gender, and Consumption: Public Discourses on Woman and the Home" (Ph.D. diss., University of Iowa, 1987).

18. Harvey Levenstein, *Revolution at the Table: The Transformation of the American Diet* (New York: Oxford University Press, 1988).

19. *Ladies' Home Journal,* February 1933, 67.

20. Pineapple advertisements highlighting nutrition in the *Ladies' Home Journal* were published in the following issues: March 1933, 83; April 1933, 114; May 1933, 83, 120; June 1933, 97; July 1933, 96; November 1933, 82; December 1933, 83, 122; January 1934, 88; February 1934, 66; March 1934, 69, 91; April 1934, 57; and May 1934, 82.

21. Gove Hambidge, "This New Era in Foods," *Ladies' Home Journal,* May 1929, 26–27, 156–57, 159. As if to prove Hambidge wrong, the National Canners Association placed an advertisement in the *Journal,* May 1934, 105, deceptively presented as an article titled "Ladies' Aid" about processed and canned foods, "food so meticulously prepared that no

suspicion of loss of nutriment or purity or wholesomeness can be laid to it. And when you think of flavor—well, here it is at its highest and best."

22. Fisher, *American Cookbook,* 52, 53, 54; and Mary E. Williams and Katharine Rolston Fisher, *Elements of the Theory and Practice of Cookery* (New York: Macmillan, 1911), which was written by a supervisor and teacher of public schools in Manhattan and the Bronx.

23. Fisher, *American Cookbook,* 126–27. Cookbooks also mirrored the debate over modernity and its stress on speed and convenience, with dissenters urging the "natural," fresh, and wholesome.

24. As quoted in James Sloan Allen, *The Romance of Commerce and Culture: Capitalism, Modernism, and the Chicago-Aspen Crusade for Cultural Reform* (Chicago: University of Chicago Press, 1983), 8.

25. Charles T. Coiner, "Atelier to Advertising," *Advertising Arts,* March 1934, 10.

26. See Beechert, *Honolulu,* 117–21.

27. DeSoto Brown, *Hawaii Recalls: Selling Romance to America* (Honolulu: Editions Limited, 1982), 86.

28. As quoted in Martin Greif, *Depression Modern: The Thirties Style in America* (New York: Universe Books, 1975), 32.

29. On women's magazines of this period, see Scanlon, *Inarticulate Longings;* Helen Damon-Moore, *Magazine for the Millions: Gender and Commerce in the* Ladies' Home Journal *and the* Saturday Evening Post, *1880–1910* (Albany: State University of New York Press, 1994); Mary Ellen Zuckerman, *A History of Popular Women's Magazines in the United States, 1792–1995* (Westport, Conn.: Greenwood Press, 1998); and Nancy A. Walker, *Shaping Our Mothers' World: American Women's Magazines* (Jackson: University Press of Mississippi, 2000).

30. Scanlon, *Inarticulate Longings,* 14.

31. As quoted from Ligon's *True and Exact History of Barbados* (1657) in Beauman, *Pineapple,* 40, 41.

32. Until independence, American foods were transplants of the English table, but outside of New England and in postcolonial America American cuisine followed a West Indian path of adaptation and negotiation among a variety of traditions of the masses and fringes of empire and not of the English or European elite of the core, according to James E. McWilliams, *A Revolution in Eating: How the Quest for Food Shaped America* (New York: Columbia University Press, 2005); and Donna R. Gabaccia, *We Are What We*

Eat: Ethnic Food and the Making of Americans (Cambridge, Mass.: Harvard University Press, 1998). Cf. Waverley Root and Richard de Rochemont, *Eating in America: A History* (New York: William Morrow, 1976), 9–11, 276, who argue that the United States was essentially English and not "a culinary melting pot."

33. Beauman, *Pineapple,* 128.

34. Beauman, *Pineapple,* 128–29, 130. Philadelphia prices in 1786 appeared reasonable to Ann Warder, an immigrant from London and member of the merchant class. Warder reported that "provisions of every kind are cheaper," with pineapple, strawberry, and cherry in abundance. "Extracts from the Diary of Mrs. Ann Warder," *Pennsylvania Magazine of History and Biography* 17:4 (1893): 448.

35. Beauman, *Pineapple,* 123–25, 132–34. On the matter of food, gastronomy, and rank and inequality, see Fernández-Armesto, *Food.* On Jefferson the epicure, see Marie Kimball, *Thomas Jefferson's Cook Book* (Charlottesville: University Press of Virginia, 1976), 1–4, 6–9, 18; "Pineapple Pudding" (p. 95) is from the Monticello book of recipes compiled by Virginia Randolph, daughter of Martha Jefferson Randolph, born in 1801. The recipe has been updated for current use.

36. Find the identical recipe in "The Housewife," *Ballou's Monthly Magazine* 23:2 (1866): 162.

37. Beauman, *Pineapple,* 138–39; and Gladys E. Bolhouse, "Abraham Redwood: Reluctant Quaker, Philanthropist, Botanist," in *Redwood Papers: A Bicentennial Collection,* ed. Lorraine Dexter and Alan Pryce-Jones (Newport, R.I.: Redwood Library and Athenaeum, 1976), 1–11.

38. American gardeners in the Boston area grew pineapples in hothouses after Redwood's success, as was reported in 1794 of Ashton Harvey of Salem and Joseph Barrell of Boston. Belden, *Festive Tradition,* 222.

39. As quoted in Beauman, *Pineapple,* 198.

40. Beauman, *Pineapple,* 199.

41. "The Pineapple," *Family Magazine* 4 (1836): 396; and "The Pineapple," *Dwights American Magazine,* June 27, 1846, 328.

42. George N. Stack, "Culture of the Pineapple," *Horticulturist and Journal of Rural Art and Rural Taste* 20 (May 1865): 151–52.

43. Henry David Thoreau, *The Journal of Henry D. Thoreau,* vol. 14, ed. Bradford Torrey and Francis H. Allen (Boston: Houghton Mifflin, 1906), 273, 274.

44. Beauman, *Pineapple,* 204, 207–8; and Hawkins, "Baltimore Canning Industry," 2, 11. For an account of blockade running from Charleston to Nassau, see "From Nassau,"

New York Times, May 24, 1863, 1; and "Running the Charleston Blockade," *New York Observer and Chronicle,* March 2, 1865, 69.

45. "Pineapple," *Dwights,* June 27, 1846, 328; "The Fruit Trade," *American Phrenological Journal* 22 (November 1855): 117; "The West India Fruit Trade," *Ohio Farmer,* July 18, 1868, 454–55; "Tropical Fruit," *Friend,* August 11, 1877, 411–12; "Pineapples," *Independent,* December 13, 1877, 30; (Mrs.) Frank Leslie, "The Pineapple Trade in the Bahamas," *Frank Leslie's Popular Monthly* 11:3 (1881): 364–66; and Hawkins, "Baltimore Canning Industry," 4–16.

46. For contemporary reports on Florida's pineapples, see J. A. Macdonald, "Pineapple Culture in Florida," *Forest and Stream,* August 19, 1875, 20; "American Pineapples," *Scientific American,* December 2, 1893, 357; Kirk Munroe, "Pineapples," *Youth's Companion,* October 26, 1893, 503–4; Wm. P. Neeld, "Pineapples in Florida," *Southern Cultivator* 3:5 (1894): 244; and James Mott, "The Pineapple," *Southern Cultivator* 52:10 (1894): 484–86.

47. See Virginia Aronson, *Konnichiwa Florida Moon: The Story of George Morikami, Pineapple Pioneer* (Sarasota, Fla.: Pineapple Press, 2002); and Kesa Noda, *Yamato Colony: 1906–1960, Livingston, California* (Livingston, Calif.: Livingston-Merced JACL Chapter, 1981). For a survey of anti-Japanese activity during this period, see Roger Daniels, *The Politics of Prejudice: The Anti-Japanese Movement in California and the Struggle for Japanese Exclusion* (New York: Atheneum, 1970).

48. M. E. Bamford, "Pineapples," *Independent,* February 23, 1888, 30–31; and George E. Walsh, "Pineapple Cultivation," *Independent,* January 31, 1889, 30.

49. "The Housewife," *Ballou's Monthly Magazine* 23:2 (1866): 162; "How to Eat Oranges and Pineapples," *Arthur's Illustrated Home Magazine* 43:8 (1875): 518–19; Katherine Armstrong, "The Pineapple," *Independent,* May 8, 1890, 39; Mabel D. Hauck, "Pineapples," *Ohio Farmer,* July 21, 1898, 45; and Eleanor M. Lucas, "What May Be Done with Pineapples," *Ladies' Home Journal,* June 1902, 34.

50. Katherine B. Johnson, "The Fragrant Pineapple," *New York Observer and Chronicle,* June 15, 1899, 793.

51. "Medicinal Virtues of the Pineapple," *Coleman's Rural World,* August 27, 1902, 6; "Medicinal Value of Pineapple Juice," *Christian Advocate,* March 26, 1903, 510; and Lewis B. Allyn, "Questions Concerning Foods and Drugs," *McClure's Magazine* 49:3 (1917): 58.

52. On the origins of crochet called "nun's work" or "nun's lace," see Lis Paludan, *Crochet: History & Technique* (Loveland, Colo.: Interweave Press, 1995).

53. Annie Louise Potter, *A Living Mystery: The International Art & History of Crochet* (n.p.: A.J. Publishing International, 1990), 41.

54. Anthea Callen, *Women Artists of the Arts and Crafts Movement, 1870–1914* (New York: Pantheon Books, 1979), 96, 97. See also Anne L. Macdonald, *No Idle Hands: The Social History of American Knitting* (New York: Ballantine Books, 1988), 175–98.

55. Paludan, *Crochet,* 41, 46, 62–66, 78; and Potter, *Living Mystery,* 101–25.

56. Potter, *Living Mystery,* 127, 129.

57. Potter, *Living Mystery,* 96.

58. Mary Gay Humphreys, "Embroidery Notes," *Art Amateur* 7:4 (1882): 85A; "Lace Work," *Ohio Farmer,* August 22, 1885, 126–27; "Pineapple Lace," *Ohio Farmer,* May 29, 1886, 366; "Handsome Crochet Edge," *Ladies' Home Journal,* July 1886, 4; "Pineapple Insertion," *Ladies' Home Journal,* July 1886, 4; "Pineapple Doily," *Ohio Farmer,* April 13, 1899, 331; and "Home Decoration and Fancy Needlework," *Arthur's Home Magazine* 61 (August 1891): 657–58.

59. Mary Gay Humphreys, "Oriental Embroidery," *Art Amateur* 4:5 (1881): 100–101.

60. "Lace-Making in America," *Art Amateur* 1:2 (1879): 39.

61. "The Art Embroidery Revival," *Art Amateur* 2:5 (1880): 103.

62. Potter, *Living Mystery,* 130–32.

63. Eugenicist Lothrop Stoddard claimed the war destroyed European supremacy and involved "all the white nations in a common ruin," leaving the gates unguarded against "the rising tide of color." Stoddard, *Rising Tide,* 208, 221.

64. Potter, *Living Mystery,* 133.

65. Collins, *Pineapple,* 26; and Jean Gorely, "The Pineapple: Symbol of Hospitality," *Antiques* 48:1 (1945): 24. Gorely also notes that from 1775 to the mid-nineteenth century the fruit symbolized "the pinnacle of perfection," as when given by lovers as a gift the pineapple conveyed the message "you are perfect." That meaning seems to recuperate the original Tupí-Guaraní name for the pineapple, *nana,* meaning "excellent fruit."

66. Beauman, *Pineapple,* 130–31. Cf. Claudia Hyles, *And the Answer Is a Pineapple: The King of Fruit in Folklore, Fabric and Food* (Burra Creek, Australia: Milner, 1998), 27, which recalls this story as a historical fact.

67. After all, science and modernity were the engines of migration. For a discrete study of India, modernity, and internal migration, see Jacob John Kattakayam, *Modernity and Migration* (Trivandrum, India: Centre for Social Research, 1981).

68. Other promptings for the Colonial Revival included the style's economy, craftsmanship, femininity, and a return to order and restraint, the virtuous life, nature and the natural. William Bertolet Rhoads, "The Colonial Revival" (Ph.D. diss., Princeton University, 1974), 376–527.

69. On the creation of this modern American nationalism, see Jacobson, *Barbarian Virtues.*

70. Mackay, "Agents of Empire," 38–57.

71. Hawai'i, the Association of Hawaiian Pineapple Canners claimed in an advertisement in the *Honolulu Advertiser* of April 30, 1925, gained from its eighty-one advertisements, which reached 12 million homes in the United States. I am grateful to DeSoto Brown for locating this advertisement.

72. "Hawaiian Agriculture," *Plough, the Loom and the Anvil* 5:5 (1852): 294, 295. In a similar vein, an author exclaims, "Can nothing be done for Jamaica, where Nature does so much, and man so little?" James Linen, "Island Sketches," *Knickerbocker* 43:4 (1854): 377.

73. See, e.g., Handy and Handy, *Native Planters.*

74. As quoted in Handy and Handy, *Native Planters,* 525–26.

75. E.S. Craighill Handy in Handy and Handy, *Native Planters,* v.

76. Mackay, "Agents of Empire," 54.

77. "Tropical Fruits," *Littell's Living Age,* March 15, 1884, 700.

78. G.E.W., "The New Agriculture of the Tropics," *Scientific American,* August 4, 1900, 67.

79. G.E.W., "New Agriculture," 67. See also D. Morris, "Tropical Fruits," *American Journal of Pharmacy* 58:9 (1886): 444–47.

80. For a useful designation of "Occidentalism" as a discourse of separations and hierarchies, see Fernando Coronil, "Beyond Occidentalism: Toward Nonimperial Geohistorical Categories," *Cultural Anthropology* 11:1 (1996): 51–87.

BIBLIOGRAPHY

ARCHIVES AND COLLECTIONS

Bernice Pauahi Bishop Museum, Honolulu, Hawai'i
Hawaiian Mission Children's Society, Honolulu, Hawai'i
Hawai'i State Archives, Honolulu, Hawai'i
Hawai'i State Library, Honolulu, Hawai'i
Special Collections, Hamilton Library, University of Hawai'i, Mānoa

PERIODICALS

Advertising Arts
American Journal of Pharmacy
American Phrenological Journal
Antiques
Art Amateur
Arthur's Home Magazine
Arthur's Illustrated Home Magazine
Ballou's Monthly Magazine
Baptist Missionary Magazine
*Bulletin of the American Geographical
    Society of New York*
Century Illustrated Magazine
Christian Advocate
Christian Observer
Coleman's Rural World
Cosmopolitan Magazine
Current Opinion

Dwights American Magazine
*The Eclectic Magazine of Foreign
    Literature*
Family Magazine
Forest and Stream
Fortune
Forum
Frank Leslie's Popular Monthly
The Friend
Gardeners' Chronicle (England)
Harper's
Honolulu Star-Bulletin
*Horticulturist and Journal of Rural
    Art and Rural Taste*
Independent
*Journal of the American Geographical
    Society of New York*

Knickerbocker
Ladies' Home Journal
Littell's Living Age
Massachusetts Ploughman
    and New England Journal
    of Agriculture
McClure's Magazine
Medical News
New McClure's
New York Observer and Chronicle
New York Times
Nippu Jiji
North American Review

Ohio Farmer
Outlook
Overland Monthly and Out
    West Magazine
Pennsylvania Magazine of History
    and Biography
Plough, the Loom and the Anvil
San Francisco Examiner
Saturday Evening Post
Saturday Review of Literature
Scientific American
Southern Cultivator
Youth's Companion

SECONDARY SOURCES

Aarim-Heriot, Najia. 2003. *Chinese Immigrants, African Americans, and Racial Anxiety in the United States, 1848–82.* Urbana: University of Illinois Press.

Abu-Lughod, Janet L. 1989. *Before European Hegemony: The World System A.D. 1250–1350.* New York: Oxford University Press.

Adamson, Alan H. 1972. *Sugar without Slaves: The Political Economy of British Guiana, 1838–1904.* New Haven, Conn.: Yale University Press.

Ai, Chung Kun. 1960. *My Seventy-Nine Years in Hawaii.* Causeway Bay, Hong Kong: Cosmorama Pictorial Publisher.

Allaire, Louis. 1997. "The Lesser Antilles before Columbus." In *The Indigenous Peoples of the Caribbean,* edited by Samuel M. Wilson, 20–28. Gainesville: University Press of Florida.

Allen, Helena G. 1988. *Sanford Ballard Dole: Hawaii's Only President, 1844–1926.* Glendale, Calif.: Arthur H. Clark.

Allen, James Sloan. 1983. *The Romance of Commerce and Culture: Capitalism, Modernism, and the Chicago-Aspen Crusade for Cultural Reform.* Chicago: University of Chicago Press.

Allen, John L. 1976. "Lands of Myth, Waters of Wonder: The Place of the Imagination in the History of Geographical Exploration." In *Geographies of the Mind: Essays in Historical Geosophy,* edited by David Lowenthal and Martyn J. Bowden, 41–61. New York: Oxford University Press.

Allyn, Lewis B. 1917. "Questions Concerning Foods and Drugs." *McClure's Magazine* 49, no. 3 (July): 58.

Altman, Karen Elizabeth. 1987. "Modernity, Gender, and Consumption: Public Discourses on Woman and the Home." Ph.D. dissertation, University of Iowa.

"The American Can Company." *Fortune,* November 1930, 39–42.

"American Pineapples." *Scientific American,* December 2, 1893, 357.

Anderson, Warwick. 1992. "Climates of Opinion: Acclimatization in Nineteenth-Century France and England." *Victorian Studies* 35, no. 2 (Winter): 135–57.

Andrade, Ernest, Jr. 1996. *Unconquerable Rebel: Robert W. Wilcox and Hawaiian Politics, 1880–1903.* Niwot: University Press of Colorado.

Aristotle. 1988. *The Politics,* translated by Benjamin Jowett. Cambridge: Cambridge University Press.

Armstrong, Katherine. 1890. "The Pineapple." *Independent,* May 8, 39.

S.C.A. [Armstrong, Samuel Chapman]. 1887. "Reminiscences." In *Richard Armstrong: America, Hawaii.* Hampton, Va.: Normal School Steam Press.

Arnold, David. 2000. " 'Illusory Riches': Representations of the Tropical World, 1840–1950." *Singapore Journal of Tropical Geography* 21, no. 1 (March): 6–18.

———. 2006. *The Tropics and the Traveling Gaze: India, Landscape, and Science, 1800–1856.* Seattle: University of Washington Press.

Aronson, Virginia. 2002. *Konnichiwa Florida Moon: The Story of George Morikami, Pineapple Pioneer.* Sarasota, Fla.: Pineapple Press.

"The Art Embroidery Revival." *Art Amateur* 2, no. 5 (April 1880): 103.

Arvelo-Jiménez, Nelly, and Horacio Biord. 1994. "The Impact of Conquest on Contemporary Indigenous Peoples of the Guiana Shield: The System of Orinoco Regional Interdependence." In *Amazonian Indians from Prehistory to the Present: Anthropological Perspectives,* edited by Anna Roosevelt, 55–78. Tucson: University of Arizona Press.

Baker, Kenneth F., and J. L. Collins. 1939. "Notes on the Distribution and Ecology of Ananas and Pseudananas in South America." *American Journal of Botany* 26, no. 9 (November): 697–702.

Balée, William. 1984. "The Ecology of Ancient Tupi Warfare." In *Warfare, Culture, and Environment,* edited by R. Brian Ferguson, 241–65. Orlando, Fla.: Academic Press.

———. 1994. *Footprints of the Forest: Ka'apor Ethnobotany: The Historical Ecology of Plant Utilization by an Amazonian People.* New York: Columbia University Press.

Bamford, M. E. 1888. "Pineapples." *Independent,* February 23, 30–31.

Barnes, Harry Elmer. 1935. "Introduction." In *The Banana Empire: A Case Study of Economic Imperialism,* edited by Charles David Kepner Jr. and Jay Henry Soothill, 3–24. New York: Vanguard Press.

Bassin, Mark. 1987. "Imperialism and the Nation State in Friedrich Ratzel's Political Geography." *Progress in Human Geography* 11, no. 4 (September): 473–95.

Baudet, Henri. 1965. *Paradise on Earth: Some Thoughts on European Images of Non-European Man,* translated by Elizabeth Wentholt. New Haven, Conn.: Yale Univesity Press.

Belden, Louise Conway. 1983. *The Festive Tradition: Table Decoration and Desserts in America, 1650–1900.* New York: W. W. Norton.

Beauman, Fran. 2005. *The Pineapple: King of Fruits.* London: Chatto and Windus.

Beckford, George L. 1983. *Persistent Poverty: Underdevelopment in Plantation Economies of the Third World.* London: Zed Books.

Bederman, Gail. 1995. *Manliness and Civilization: A Cultural History of Gender and Race in the United States, 1880–1917.* Chicago: University of Chicago Press.

Beechert, Edward D. 1985. *Working in Hawaii: A Labor History.* Honolulu: University of Hawai'i Press.

———. 1991. *Honolulu: Crossroads of the Pacific.* Columbia: University of South Carolina Press.

Begley, Vimala, and Richard Daniel De Puma, eds. 1991. *Rome and India: The Ancient Sea Trade.* Madison: University of Wisconsin Press.

Bernstein, Richard, and Ross H. Munro. 1997. *The Coming Conflict with China.* New York: Alfred A. Knopf.

Black, Edwin. 2003. *War against the Weak: Eugenics and America's Campaign to Create a Master Race.* New York: Four Walls Eight Windows.

Blumenbach, Johann Friedrich. 1969. *On the Natural Varieties of Mankind,* edited by Thomas Bendyshe. New York: Bergman Publishers.

Bodin, John. 1945. *Method for the Easy Comprehension of History,* translated by Beatrice Reynolds. New York: Columbia University Press.

Bolhouse, Gladys E. 1976. "Abraham Redwood: Reluctant Quaker, Philanthropist, Botanist." In *Redwood Papers: A Bicentennial Collection,* edited by Lorraine Dexter and Alan Pryce-Jones, 1–11. Newport, R.I.: Redwood Library and Athenaeum.

Boothe, Clare. 1942. "Ever Hear of Homer Lea?" *Saturday Evening Post,* March 7, 12–13, 69–72; and March 14, 27, 38–40, 42.

Boxer, C. R. 1957. *The Dutch in Brazil, 1624–1654.* Oxford, England: Clarendon Press.

Brading, D. A. 1991. *The First America: The Spanish Monarchy, Creole Patriots, and the Liberal State, 1492–1867.* Cambridge: Cambridge University Press.

Braudel, Fernand. 1992. *Civilization and Capitalism, 15th–18th Century.* Vol. 3, *The Perspective of the World,* translated by Siân Reynolds. Berkeley: University of California Press.

Brockway, Lucile H. 1979. *Science and Colonial Expansion: The Role of the British Royal Botanic Gardens.* New York: Academic Press.

Brown, DeSoto. 1982. *Hawaii Recalls: Selling Romance to America.* Honolulu: Editions Limited.

Bryce, James. 1892. "The Migrations of the Races of Men Considered Historically." *Eclectic Magazine of Foreign Literature* 56, no. 3 (September): 289–303.

———. 1899. "British Experience in the Government of Colonies." *Century Illustrated Magazine* 62, no. 5 (March): 718–28.

Bunbury, E. H. 1959. *A History of Ancient Geography.* Vol. 1, 2d ed. New York: Dover.

Burke, Donald S. 2003. "American Society of Tropical Medicine and Hygiene: Centennial Celebration Address." December 3. www.astmh.org/address.pdf.

Burton, Wilbur. 1942. "Prophet of Pearl Harbor." *Saturday Review of Literature* 25, no. 14 (April 4): 8.

Callen, Anthea. 1979. *Women Artists of the Arts and Crafts Movement, 1870–1914.* New York: Pantheon Books.

Campbell, Mary B. 1988. *The Witness and the Other World: Exotic European Travel Writing, 400–1600.* Ithaca, N.Y.: Cornell University Press.

"Canned Pineapple Recipes." *Christian Observer,* March 30, 1910, 18.

Carson, Anne. 1990. "Putting Her in Her Place: Woman, Dirt, and Desire." In *Before Sexuality: The Construction of Erotic Experience in the Ancient Greek World,* edited by David M. Halperin, John J. Winkler, and Froma I. Zeitlin, 134–69. Princeton, N.J.: Princeton University Press.

Carter, Harold B. 1988. *Sir Joseph Banks, 1743–1820.* London: British Museum.

Carter, Paul. 1987. *The Road to Botany Bay: An Essay in Spatial History.* London: Faber and Faber.

Cartoons of the War of 1898 with Spain from Leading Foreign and American Papers. Chicago: Belford, Middlebrook, 1898.

Chamberlain, Mary Endicott. 1900. "An Obligation of Empire." *North American Review* 170, no. 71 (April): 493–503.

Chan, Sucheng. 1986. *This Bitter-Sweet Soil: The Chinese in California Agriculture, 1860–1910.* Berkeley: University of California Press.

Chapman, Royal N. 1933. *Cooperation in the Hawaiian Pineapple Business.* New York: Institute for Pacific Relations.

Char, Tin-Yuke, ed. 1975. *The Sandalwood Mountains: Readings and Stories of the Early Chinese in Hawaii.* Honolulu: University of Hawai'i Press.

Chinen, Jon J. 1958. *The Great Mahele: Hawaii's Land Division of 1848.* Honolulu: University of Hawai'i Press.

Clarke, Marcia. 1930. "Tinned Food for Tin Gods." *Forum,* February, 94–96.

Clarke, Thurston. 1991. *Pearl Harbor Ghosts: A Journey to Hawaii Then and Now.* New York: William Morrow.

Coan, Titus. 1882. *Life in Hawaii: An Autobiographical Sketch of Mission Life and Labors, 1835–1882.* New York: Anson D. F. Randolph.

Coiner, Charles T. 1934. "Atelier to Advertising." *Advertising Arts,* March, 9–10.

Collins, J. L. 1948. "Pineapples in Ancient America." *Scientific Monthly* 67, no. 5 (November): 372–77.

———. 1951. "Notes on the Origin, History, and Genetic Nature of the Cayenne Pineapple." *Pacific Science* 5, no. 1 (January): 3–17.

———. 1960. *The Pineapple: Botany, Cultivation, and Utilization.* London: Leonard Hill.

[Columbus, Christopher]. 1987. *The Log of Christopher Columbus,* translated by Robert H. Fuson. Camden, Me.: International Marine Publishing.

Conroy, Hilary. 1953. *The Japanese Frontier in Hawaii, 1868–1898.* Berkeley: University of California Press.

Cook, Warren L. 1973. *Flood Tide of Empire: Spain and the Pacific Northwest, 1543–1819.* New Haven, Conn.: Yale University Press.

Cooke, Ian, et al. 2004. "Follow the Thing: Papaya." *Antipode* 36, no. 4: 642–64.

Cooper, George, and Gavan Daws. 1985. *Land and Power in Hawaii: The Democratic Years.* Honolulu: Benchmark Books.

Coronil, Fernando. 1996. "Beyond Occidentalism: Toward Nonimperial Geohistorical Categories." *Cultural Anthropology* 11, no. 1 (February): 51–87.

Corrêa do Lago, Pedro, and Blaise Ducos, eds. 2005. *Frans Post: Le Brésil à la cour de Louis XIV.* Milan: 5 Continents Edition.

Crichton, Michael. 1992. *Rising Sun.* New York: Ballantine Books.

Crosby, Alfred W., Jr. 1972. *The Columbian Exchange: Biological and Cultural Consequences of 1492.* Westport, Conn.: Greenwood Press.

Curtin, Philip D. 1989. *Death by Migration: Europe's Encounter with the Tropical World in the Nineteenth Century.* Cambridge: Cambridge University Press.

———. 1990. *The Rise and Fall of the Plantation Complex.* Cambridge: Cambridge University Press.

Cutter, Donald. 1980. "The Spanish in Hawaii: Gaytan to Marin." *Hawaiian Journal of History* 14: 16–25.

Damon-Moore, Helen. 1994. *Magazine for the Millions: Gender and Commerce in the Ladies' Home Journal and the Saturday Evening Post, 1880–1910.* Albany: State University of New York Press.

"Dangers of Canned Food." *Medical News,* September 8, 1883, 270–71.

Daniels, Roger. 1970. *The Politics of Prejudice: The Anti-Japanese Movement in California and the Struggle for Japanese Exclusion*. New York: Atheneum.

Daws, Gavan. 1968. *Shoal of Time: A History of the Hawaiian Islands*. Honolulu: University of Hawai'i Press.

Dean-Jones, Leslie. 1994. *Women's Bodies in Classical Greek Science*. Oxford, England: Clarendon Press.

Delaporte, François. 1991. *The History of Yellow Fever: An Essay on the Birth of Tropical Medicine*, translated by Arthur Goldhammer. Cambridge, Mass.: MIT Press.

De Sousa-Leão, Joaquim. 1973. *Frans Post, 1612–1680*. Amsterdam: A. L. Van Gendt.

Desowitz, Robert S. 1997. *Who Gave Pinta to the Santa Maria? Torrid Diseases in a Temperate World*. New York: W. W. Norton.

Dettelbach, Michael. 1996. "Global Physics and Aesthetic Empire: Humboldt's Physical Portrait of the Tropics." In *Visions of Empire: Voyages, Botany, and Representations of Nature*, edited by David Philip Miller and Peter Hanns Reill, 258–92. Cambridge: Cambridge University Press.

"Diseases of Hot Climates." *Medical News* 83, no. 2 (July 11, 1903): 84–86.

Dole, Richard, and Elizabeth Dole Porteus. 1990. *The Story of James Dole*. 'Aiea, Hawai'i: Island Heritage Publishing.

Dooner, Pierton W. 1879. *Last Days of the Republic*. San Francisco: Alta California Publishing.

Dower, John W. 1986. *War without Mercy: Race and Power in the Pacific War*. New York: Pantheon.

Du Bois, W. E. Burghardt. 1956. *Black Reconstruction in America*. New York: Russell and Russell.

Duffy, John. 1990. *The Sanitarians: A History of American Public Health*. Urbana: University of Illinois Press.

"Effects of Heat and Cold on Canned Foods." *Scientific American*, May 27, 1893, 325.

Elliott, J. H. 1970. *The Old World and the New, 1492–1650*. Cambridge: Cambridge University Press.

———. 1984. "The Spanish Conquest and Settlement of America." In *The Cambridge History of Latin America*. Vol. 1, *Colonial Latin America*, edited by Leslie Bethell, 149–62. Cambridge: Cambridge University Press.

Erickson, Ethel. 1940. *Earnings and Hours in Hawaii Woman-Employing Industries*. Bulletin No. 177. Women's Bureau, U.S. Department of Labor. Washington, D.C.: Government Printing Office.

Estes, J. Worth, and Billy G. Smith, eds. 1997. *A Melancholy Scene of Devastation: The Public*

Response to the 1793 Philadelphia Yellow Fever Epidemic. Canton, Mass.: Science History Publications.

Eustis, Elizabeth S. 2003. *European Pleasure Gardens: Rare Books and Prints of Historic Landscape Design from the Elizabeth K. Reilley Collection.* New York: New York Botanical Garden.

Ewen, Stuart. 1976. *Captains of Consciousness.* New York: McGraw Hill.

Febvre, Lucien. 1925. *A Geographical Introduction to History,* translated by E. G. Mountford and J. H. Paxton. New York: Alfred A. Knopf.

Felkin, R. W. 1889. *On the Geographical Distribution of Some Tropical Diseases, and Their Relation to Physical Phenomena.* Edinburgh, Scotland: Young J. Pentland.

Fernández-Armesto, Felipe. 2001. *Food: A History.* London: Macmillan.

Fisher, Carol. 2006. *The American Cookbook: A History.* Jefferson, N.C.: McFarland and Company.

Flint, Valerie I. J. 1992. *The Imaginative Landscape of Christopher Columbus.* Princeton, N.J.: Princeton University Press.

Frank, Andre Gunder. 1967. *Capitalism and Underdevelopment in Latin America: Historical Studies of Chile and Brazil.* New York: Monthly Review Press.

———. 1998. *ReOrient: Global Economy in the Asian Age.* Berkeley: University of California Press.

Frederick, Christine. 1929. *Selling Mrs. Consumer.* New York: Business Bourse.

Freedman, Paul, ed. 2007. *Food: A History of Taste.* Berkeley: University of California Press.

Friedman, George, and Meredith Lebard. 1991. *The Coming War with Japan.* New York: St. Martin's Press.

Frost, Alan. 1996. "The Antipodean Exchange: European Horticulture and Imperial Designs." In *Visions of Empire: Voyages, Botany, and Representations of Nature,* edited by David Philip Miller and Peter Hanns Reill, 58–79. Cambridge: Cambridge University Press.

"Fruit Canning." *Massachusetts Ploughman and New England Journal of Agriculture,* January 15, 1870, 1.

"The Fruit Trade." *American Phrenological Journal* 22 (November 1855): 117.

Fuchs, Lawrence H. 1961. *Hawaii Pono: A Social History.* New York: Harcourt, Brace and World.

Fujitani, T., Geoffrey M. White, and Lisa Yoneyama, eds. 2001. *Perilous Memories: The Asia-Pacific War(s).* Durham, N.C.: Duke University Press.

Fuller, Mary C. 1993. "Raleigh's Fugitive Gold: Reference and Deferral in *The Discoverie of Guiana.*" In *New World Encounters,* edited by Stephen Greenblatt, 218–40. Berkeley: University of California Press.

Gabaccia, Donna R. 1998. *We Are What We Eat: Ethnic Food and the Making of Americans.* Cambridge, Mass.: Harvard University Press.

Gallicchio, Marc. 2000. *The African American Encounter with Japan and China: Black Internationalism in Asia, 1895–1945.* Chapel Hill: University of North Carolina Press.

Galloway, J. H. 1989. *The Sugar Cane Industry: An Historical Geography from Its Origins to 1914.* Cambridge: Cambridge University Press.

Gast, Ross H. 1975. *Don Francisco de Paula Marin: A Biography.* Honolulu: University of Hawai'i Press.

Gay, Lawrence Kainoahou. 1965. *True Stories of the Island of Lanai.* Honolulu: Mission Press.

G. E. W. 1900. "The New Agriculture of the Tropics." *Scientific American,* August 4, 67.

Gibson, Arrell Morgan. 1993. *Yankees in Paradise: The Pacific Basin Frontier.* Albuquerque: University of New Mexico Press.

Glacken, Clarence J. 1967. *Traces on the Rhodian Shore: Nature and Culture in Western Thought from Ancient Times to the End of the Eighteenth Century.* Berkeley: University of California Press.

Glick, Clarence E. 1980. *Sojourners and Settlers: Chinese Migrants in Hawaii.* Honolulu: Hawaii Chinese History Center.

Gobineau, Arthur de. 1967. *The Inequality of Human Races,* translated by Adrian Collins. New York: Howard Fertig.

Godlewska, Anne, and Neil Smith, eds. 1994. *Geography and Empire.* Oxford, England: Blackwell.

Gorely, Jean. 1945. "The Pineapple: Symbol of Hospitality." *Antiques* 48, no. 1 (July): 22–24.

Gossett, Thomas F. 1997. *Race: The History of an Idea in America.* New York: Oxford University Press.

Gould, Stephen Jay. 1981. *The Mismeasure of Man.* New York: W. W. Norton.

Grant, Madison. 1920. "Introduction." In *The Rising Tide of Color against White World-Supremacy,* by Lothrop Stoddard, xi–xxxii. New York: Charles Scribner's Sons.

Greenbie, Sidney, and Marjorie Barstow. 1937. *Gold of Ophir: The China Trade in the Making of America.* New York: Wilson-Erickson.

Greenblatt, Stephen. 1991. *Marvelous Possessions: The Wonder of the New World.* Chicago: University of Chicago Press.

———, ed. 1993. *New World Encounters.* Berkeley: University of California Press.

Gregory, J. W. 1925. *The Menace of Colour: A Study of the Difficulties Due to the Association of White & Coloured Races, with an Account of Measures Proposed for Their Solution, & Special Reference to White Colonization in the Tropics.* London: Seeley, Service.

Greif, Martin. 1975. *Depression Modern: The Thirties Style in America.* New York: Universe Books.

Grenier, Josephine. 1909. "New Uses for Canned Fruits." *Harper's* 43, no. 3 (March): 266–68.

Gussow, Zachary. 1989. *Leprosy, Racism, and Public Health: Social Policy in Chronic Disease Control.* Boulder, Colo.: Westview Press.

Guyot, Arnold Henry. 1849. *The Earth and Man: Lectures on Comparative Physical Geography in Its Relation to the History of Mankind.* New York: Sheldon, Blakeman.

Hall, Prescott F. 1906. *Immigration and Its Effects upon the United States.* New York: Henry Holt.

———. 1919. "Immigration Restriction and World Eugenics," *Journal of Heredity* 10, no. 3 (March): 125–27.

Hambidge, Gove. 1929. "This New Era in Foods." *Ladies' Home Journal,* May, 26–27, 156–57, 159.

Handlin, Oscar. 1951. *The Uprooted: The Epic Story of the Great Migrations That Made the American People.* New York: Grosset and Dunlop.

"Handsome Crochet Edge." *Ladies' Home Journal,* July 1886, 4.

Handy, E. S. Craighill, and Elizabeth Green Handy. 1972. *Native Planters in Old Hawaii: Their Life, Lore and Environment.* Bulletin 233. Honolulu: Bishop Museum Press.

Hannaford, Ivan. 1996. *Race: The History of an Idea in the West.* Washington, D.C.: Woodrow Wilson Center Press.

Hansen, Marcus Lee. 1940. *The Atlantic Migration, 1607–1860: A History of the Continuing Settlement of the United States,* edited by Arthur M. Schlesinger. Cambridge, Mass.: Harvard University Press.

Harris, Neil. 1985. "Designs on Demand: Art and the Modern Corporation." In *Art, Design, and the Modern Corporation.* Washington, D.C.: Smithsonian Institution Press.

Harris, Paul William. 1999. *Nothing but Christ: Rufus Anderson and the Ideology of Protestant Foreign Missions.* New York: Oxford University Press.

Harrison, Mark. 1999. *Climates and Constitutions: Health, Race, Environment and British Imperialism in India, 1600–1850.* New Delhi, India: Oxford University Press.

"The Harvard School of Tropical Medicine." *Outlook,* October 18, 1913, 343.

Hauck, Mabel D. 1898. "Pineapples." *Ohio Farmer,* July 21, 45.

"Hawaiian Agriculture." *Plough, the Loom and the Anvil* 5, no. 5 (November 1852): 294–95.

Hawaiian Pineapple as 100 Good Cooks Serve It. San Francisco: Association of Hawaiian Pineapple Canners, 1928.

Hawkins, Richard. 1995. "The Baltimore Canning Industry and the Bahamian Pineapple Trade, c. 1865–1926." *Maryland Historian* 26, no. 2 (Fall/Winter): 1–22.

Hegel, G. W. F. 1881. *Lectures on the Philosophy of History,* translated by J. Sibree. London: George Bell and Sons.

Helms, Mary W. 1978. "The Indians of the Caribbbean and Circum-Caribbean at the

End of the Fifteenth Century." In *The Cambridge History of Latin America*. Vol. 1, *Colonial Latin America*, edited by Leslie Bethell, 37–57. Cambridge: Cambridge University Press.

Hemming, John. 1978. *Red Gold: The Conquest of the Brazilian Indians*. Cambridge, Mass.: Harvard University Press.

———. 1984. "The Indians of Brazil in 1500." In *The Cambridge History of Latin America*. Vol. 1, *Colonial Latin America*, edited by Leslie Bethell, 119–46. Cambridge: Cambridge University Press.

Hidalgo, Jorge. 1984. "The Indians of Southern South America in the Middle of the Sixteenth Century." In *The Cambridge History of Latin America*. Vol. 1, *Colonial Latin America*, edited by Leslie Bethell, 91–117. Cambridge: Cambridge University Press.

Himmelfarb, Gertrude. 1959. *Darwin and the Darwinian Revolution*. New York: W. W. Norton.

[Hippocrates]. 1923. *Hippocrates*. Vol. 1, translated by W. H. S. Jones. Cambridge, Mass.: Harvard University Press.

Hobsbawm, E. J. 1975. *The Age of Capital, 1848–1875*. London: Weidenfeld and Nicolson.

Hobson, Richmond Pearson. 1908. *Cosmopolitan Magazine* 54, no. 6 (May): 584–93; and 55, no. 1 (June): 38–47.

Hoganson, Kristin L. 1998. *Fighting for American Manhood: How Gender Politics Provoked the Spanish-American and Philippine-American Wars*. New Haven, Conn.: Yale University Press.

Holt, Elizabeth Mary. 2002. *Colonizing Filipinas: Nineteenth-Century Representations of the Philippines in Western Historiography*. Manila, Philippines: Ateneo de Manila University Press.

"Home Decoration and Fancy Needlework." *Arthur's Home Magazine* 61 (August 1891): 657–58.

Horne, Gerald. 2004. *Race War! White Supremacy and the Japanese Attack on the British Empire*. New York: New York University Press.

"The Housewife." *Ballou's Monthly Magazine* 23, no. 2 (February 1866): 162.

"How to Eat Oranges and Pineapples." *Arthur's Illustrated Home Magazine* 43, no. 8 (August 1875): 518–19.

Hudson, Brian. 1977. "The New Geography and the New Imperialism: 1870–1918." *Antipode* 9, no. 2: 12–19.

Humboldt, Alexander von. 1836–39. *Examen critique de l'histoire de la géographie du Nouveau Continent et des progrès de l'astronomie nautique aux quinzième et seizième siècles*. Vols. 1–5. Paris.

———. 1889. *Personal Narrative of Travels to the Equinoctial Regions of America, during the Years 1799–1804.* Vol. 1, translated and edited by Thomasina Ross. London: George Bell and Sons.

Humphreys, Mary Gay. 1881. "Oriental Embroidery." *Art Amateur* 4, no. 5 (April): 100–101.

———. 1882. "Embroidery Notes." *Art Amateur* 7, no. 4 (September): 85A.

Huntington, Ellsworth. 1915. *Civilization and Climate.* New Haven, Conn.: Yale University Press.

———. 1924. *The Character of Races: As Influenced by Physical Environment, Natural Selection and Historical Development.* New York: Charles Scribner's Sons.

———. 1924. "Environment and Racial Character." In *Organic Adaptation to Environment,* edited by Malcolm Rutherford Thorpe, 281–99. New Haven, Conn.: Yale University Press.

Huntington, Samuel P. 1996. *The Clash of Civilizations and the Remaking of World Order.* New York: Simon and Schuster.

Huyssen, Andreas. 1986. *After the Great Divide: Modernism, Mass Culture, Postmodernism.* Bloomington: Indiana University Press.

Hyam, Ronald. 1990. *Empire and Sexuality: The British Experience.* Manchester, England: Manchester University Press.

Hyles, Claudia. 1998. *And the Answer Is a Pineapple: The King of Fruit in Folklore, Fabric and Food.* Burra Creek, Australia: Milner Publishing.

"An Industrial Empire Builder." *Outlook,* April 26, 1916, 990–92.

Inkersley, Arthur. 1911. "An American School of Tropical Medicine." *Overland Monthly and Out West Magazine* 57, no. 4 (April): 373–75.

Isaac, Allan Punzalan. 2006. *American Tropics: Articulating Filipino America.* Minneapolis: University of Minnesota Press.

Jacobson, Matthew Frye. 2000. *Barbarian Virtues: The United States Encounters Foreign Peoples at Home and Abroad, 1876–1917.* New York: Hill and Wang.

Johnson, James. 1826. *The Influence of Tropical Climates on European Constitutions: Being a Treatise on the Principal Diseases Incidental to Europeans in the East and West Indies, Mediterranean, and Coast of Africa.* New York: Duyckinck, Long, Collins and Company.

Johnson, Katherine B. 1899. "The Fragrant Pineapple." *New York Observer and Chronicle,* June 15, 793–94.

Jones, Absalom, and Richard Allen. 1794. *A Narrative of the Proceedings of the Black People, during the Late Awful Calamity in Philadelphia, in the Year 1793: and a Refutation of Some Censures, Thrown upon Them in Some Late Publications.* Philadelphia: William W. Woodward.

Jung, Moon-Ho. 2006. *Coolies and Cane: Race, Labor, and Sugar in the Age of Emancipation.* Baltimore, Md.: Johns Hopkins University Press.

Jung, Moon-Kie. 2006. *Reworking Race: The Making of Hawaii's Interracial Labor Movement.* New York: Columbia University Press.

Kamakau, Samuel M. 1992. *Ruling Chiefs of Hawaii.* Honolulu: Kamehameha Schools Press.

Kameʻeleihiwa, Lilikala. 1992. *Native Land and Foreign Desires.* Honolulu: Bishop Museum Press.

Kant, Immanuel. 1997. *Physical Geography.* In *Race and the Enlightenment: A Reader,* edited by Emmanuel Chukwudi Eze. Cambridge, Mass.: Blackwell.

Kaopuiki, Elaine Kauwenaole, and Randolph Jordan Moore. 1987. *Lanaʻi: The Mystery Island.* Honolulu: Lopa Publishing.

Kattakayam, Jacob John. 1981. *Modernity and Migration.* Trivandrum, India: Centre for Social Research.

Kearney, Reginald. 1998. *African American Views of the Japanese: Solidarity or Sedition?* Albany: State University of New York Press.

Kemp, Martin. 1996. " 'Implanted in our Natures': Humans, Plants, and the Stories of Art," In *Visions of Empire: Voyages, Botany, and Representations of Nature,* edited by David Philip Miller and Peter Hanns Reill, 207–9. Cambridge: Cambridge University Press.

Kent, Noel J. 1983. *Hawaii: Islands under the Influence.* New York: Monthly Review Press.

Kepner, Charles David, Jr., and Jay Henry Soothill. 1935. *The Banana Empire: A Case Study of Economic Imperialism.* New York: Vanguard Press.

Kidd, Benjamin. 1894. *Social Evolution.* New York: Macmillan.

———. 1898. *The Control of the Tropics.* New York: Macmillan.

Kidwell, John. 1904. "The Cultivation of Pineapples in Hawaii." *Hawaiian Forester and Agriculturist* 1, no. 12 (December): 334–45.

Kimball, Marie. 1976. *Thomas Jefferson's Cook Book.* Charlottesville: University Press of Virginia.

King, Helen. 1998. *Hippocrates' Woman: Reading the Female Body in Ancient Greece.* London: Routledge.

King, Samuel P., and Randall W. Roth. 2006. *Broken Trust: Greed, Mismanagement, and Political Manipulation at America's Largest Charitable Trust.* Honolulu: University of Hawaiʻi Press.

[Kipling, Rudyard]. 1927. *Rudyard Kipling's Verse, 1885–1926.* Garden City, New York: Doubleday, Page and Company.

Kirch, Patrick Vinton. 1985. *Feathered Gods and Fishhooks: An Introduction to Hawaiian Archaeology and Prehistory.* Honolulu: University of Hawaiʻi Press.

Kolodny, Annette. 1975. *The Lay of the Land: Metaphor as Experience and History in American Life and Letters.* Chapel Hill: University of North Carolina Press.

Kraut, Alan. 1994. *Silent Travelers: Germs, Genes and the "Immigrant Menace."* New York: HarperCollins.

Kramer, Paul A. 2003. "Empires, Exceptions, and Anglo-Saxons: Race and Rule between the British and U.S. Empires, 1880–1910." In *The American Colonial State in the Philippines: Global Perspectives,* edited by Julian Go and Anne L. Foster, 43–91. Durham, N.C.: Duke University Press.

Krempel, León, ed. 2006. *Frans Post (1612–1680): Painter of Paradise Lost.* Petersberg, Germany: Michael Imhof Verlag.

Kupperman, Karen Ordahl, ed. 1995. *America in European Consciousness, 1493–1750.* Chapel Hill: University of North Carolina Press.

Kuykendall, Ralph S. 1938. *The Hawaiian Kingdom.* Vol. 1, *Foundation and Transformation, 1778–1854.* Honolulu: University of Hawai'i Press.

———. 1953. *The Hawaiian Kingdom.* Vol. 2, *Twenty Critical Years, 1854–1874.* Honolulu: University of Hawai'i Press.

———. 1967. *The Hawaiian Kingdom.* Vol. 3, *The Kalakaua Dynasty, 1874–1893.* Honolulu: University of Hawai'i Press.

"Lace-Making in America." *Art Amateur* 1, no. 2 (July 1879): 39.

"Lace Work." *Ohio Farmer,* August 22, 1885, 126–27.

Lach, Donald F. 1965. *Asia in the Making of Europe.* Vol. 1, *The Century of Discovery.* Chicago: University of Chicago Press.

Lai, Walton Look. 1993. *Indentured Labor, Caribbean Sugar: Chinese and Indian Migrants to the British West Indies, 1838–1918.* Baltimore, Md.: Johns Hopkins University Press.

Lal, Brij V., Doug Munro, and Edward D. Beechert, eds. 1993. *Plantation Workers: Resistance and Accommodation.* Honolulu: University of Hawai'i Press.

Lapsansky, Phillip. 1997. " 'Abigail, a Negress': The Role and the Legacy of African Americans in the Yellow Fever Epidemic." In *A Melancholy Scene of Devastation: The Public Response to the 1793 Philadelphia Yellow Fever Epidemic,* edited by J. Worth Estes and Billy G. Smith, 61–78. Canton, Mass.: Science History Publications.

Larsen, Erik. 1962. *Frans Post: Interprète du Brésil.* Amsterdam: Colibris.

Laufer, Berthold. 1929. "The American Plant Migration." *Scientific Monthly* 28, no. 3 (March): 239–51.

Lavallée, Danièle. 2000. *The First South Americans: The Peopling of a Continent from the Earliest Evidence to High Culture,* translated by Paul G. Bahn. Salt Lake City: University of Utah Press.

Lea, Homer. 1909. *The Valor of Ignorance.* New York: Harper and Brothers.

Lee, Blanche Kaualua L. 2002. *A History of the Events in the Life of Hawaii's Horticulturist: Don Francisco de Paula Marin.* Honolulu: Best Printing.

Léry, Jean de [1580]. 1990. *History of a Voyage to the Land of Brazil, Otherwise Called America,* translated by Janet Whatley. Berkeley: University of California Press.

Leslie, (Mrs.) Frank. 1881. "The Pineapple Trade in the Bahamas." *Frank Leslie's Popular Monthly* 11, no. 3 (March): 364–66

Levenstein, Harvey. 1988. *Revolution at the Table: The Transformation of the American Diet.* New York: Oxford University Press.

Lévi-Strauss, Claude [1955]. 1973. *Tristes tropiques,* translated by John and Doreen Weightman. New York: Penguin Books.

Levy, Neil M. 1975. "Native Hawaiian Land Rights." *California Law Review* 63, no. 4 (July): 848–85.

Liliuokalani. 1964. *Hawaii's Story by Hawaii's Queen.* Rutland, Vt.: Charles E. Tuttle.

Linen, James. 1854. "Island Sketches." *Knickerbocker* 43, no. 4 (April): 374–77.

Livingstone, David N. 1987. "Human Acclimatization: Perspectives on a Contested Field of Inquiry in Science, Medicine and Geography." *History of Science* 25, no. 4: 359–94.

———. 1991. "The Moral Discourse of Climate: Historical Considerations on Race, Place and Virtue." *Journal of Historical Geography* 17, no. 4: 413–34.

———. 1994. "Climate's Moral Economy: Science, Race and Place in Post-Darwinian British and American Geography." In *Geography and Empire,* edited by Anne Godlewska and Neil Smith, 132–54. Oxford, England: Blackwell.

———. 2000. "Tropical Hermeneutics: Fragments for a Historical Narrative, an Afterword." *Singapore Journal of Tropical Geography* 21, no. 1 (March): 92–98.

Lucas, Eleanor M. 1902. "What May Be Done with Pineapples." *Ladies' Home Journal,* June, 34.

MacCulloch, Campbell. 1929. "The Man Who Made Hawaii." *New McClure's* 62, no. 3 (March): 20–23, 66.

Macdonald, Anne L. 1988. *No Idle Hands: The Social History of American Knitting.* New York: Ballantine Books.

Macdonald, J. A. 1875. "Pineapple Culture in Florida." *Forest and Stream,* August 19, 20.

Mackay, David. 1985. *In the Wake of Cook: Exploration, Science & Empire, 1780–1801.* New York: St. Martin's Press.

———. 1996. "Agents of Empire: The Banksian Collectors and Evaluation of New Lands." In *Visions of Empire: Voyages, Botany, and Representations of Nature,* edited by David Philip Miller and Peter Hanns Reill, 38–57. Cambridge: Cambridge University Press.

MacLeod, Roy, and Milton Lewis, eds. 1988. *Disease, Medicine, and Empire: Perspectives on Western Medicine and the Experience of European Expansion.* London: Routledge.

[Mandeville, John]. 1900. *The Travels of Sir John Mandeville*. London: Macmillan.

Manning, Caroline. 1930. *The Employment of Women in the Pineapple Canneries of Hawaii.* Bulletin no. 82. Women's Bureau, U.S. Department of Labor. Washington, D.C.: Government Printing Office.

Marchand, Roland. 1985. *Advertising the American Dream: Making Way for Modernity, 1920–1940*. Berkeley: University of California Press.

Maxon, Hazel Carter. 1927. "A Deserted Island That Became a Pineapple Plantation." *Overland Monthly and Out West Magazine* 85, no. 10 (October): 296–98.

McClintock, Anne. 1995. *Imperial Leather: Race, Gender, and Sexuality in the Colonial Contest*. New York: Routledge.

McFeely, Mary Drake. 2000. *Can She Bake a Cherry Pie? American Women and the Kitchen in the Twentieth Century*. Amherst: University of Massachusetts Press.

McGregor, Davianna Pomaikaʻi. 1996. "The Cultural and Political History of Hawaiian Native People." In *Our History Our Way: An Ethnic Studies Anthology*, edited by Gregory Yee Mark, Davianna Pomaikaʻi McGregor, and Linda A. Revilla, 333–81. Dubuque, Iowa: Kendall/Hunt.

McWilliams, James E. 2005. *A Revolution in Eating: How the Quest for Food Shaped America*. New York: Columbia University Press.

"Medical Science and the Tropics." *Bulletin of the American Geographical Society of New York* 45, no. 1 (1913): 435–38.

"Medicinal Value of Pineapple Juice." *Christian Advocate,* March 26, 1903, 510.

"Medicinal Virtues of the Pineapple." *Coleman's Rural World,* August 27, 1902, 6.

Menton, Linda K., ed. 1990. *Pineapple in Hawaii: A Guide to Historical Resources*. Honolulu: Humanities Program of the State Foundation on Culture and the Arts.

Miller, David Philip. 1996. "Joseph Banks, Empire, and 'Centers of Calculation' in Late Hanoverian London." In *Visions of Empire: Voyages, Botany, and Representations of Nature,* edited by David Philip Miller and Peter Hanns Reill, 21–37. Cambridge: Cambridge University Press.

Miller, Roger. 1991. "*Selling Mrs. Consumer:* Advertising and the Creation of Suburban Socio-Spatial Relations, 1910–1930." *Antipode* 23, no. 3: 263–306.

Milles, Ida Kanekoa. 1984. "Getting Somewheres." In *Hanahana: An Oral History Anthology of Hawaii's Working People,* edited by Michi Kodama-Nishimoto, Warren S. Nishimoto, and Cynthia A. Oshiro, 3–15. Honolulu: Ethnic Studies Oral History Project, University of Hawaiʻi, Mānoa.

Mintz, Sidney W. 1985. *Sweetness and Power: The Place of Sugar in Modern History*. New York: Viking Press.

Mohr, James C. 2005. *Plague and Fire: Battling Black Death and the 1900 Burning of Honolulu's Chinatown*. New York: Oxford University Press.

Montrose, Louis. 1993. "The Work of Gender in the Discourse of Discovery." In *New World Encounters*, edited by Stephen Greenblatt, 177–217. Berkeley: University of California Press.

Morgan, Theodore. 1948. *Hawaii, A Century of Economic Change, 1778–1876*. Cambridge, Mass.: Harvard University Press.

Moriyama, Alan Takeo. 1985. *Imingaisha: Japanese Emigration Companies and Hawaii, 1894–1908*. Honolulu: University of Hawai'i Press.

Morris, D. 1886. "Tropical Fruits." *American Journal of Pharmacy* 58, no. 9 (September): 444–47.

Mosely, Benjamin. 1787. *A Treatise on Tropical Diseases; on Military Operations; and on the Climate of the West Indies*. London.

Mott, James. 1894. "The Pineapple." *Southern Cultivator* 52, no. 10 (October): 484–86.

Munro, Doug. 1993. "Patterns of Resistance and Accommodation," in *Plantation Workers: Resistance and Accommodation*, edited by Brij V. Lal, Doug Munro, and Edward D. Beechert, 1–43. Honolulu: University of Hawai'i Press.

Munroe, Kirk. 1893. "Pineapples." *Youth's Companion*, October 26, 503–4.

Naether, Carl. 1928. *Advertising to Women*. New York: Prentice Hall.

Nash, Gary B., and Julie Roy Jeffrey. 1998. *The American People: Creating a Nation and a Society*, 4th ed. New York: Longman.

National Canners Association. 1940. *The Story of the Canning Industry*. Washington, D.C.: National Canners Association.

Navarro, Vicente, ed. 1982. *Imperialism, Health and Medicine*. London: Pluto Press.

Neeld, Wm. P. 1894. "Pineapples in Florida." *Southern Cultivator* 3, no. 5 (May): 244.

Nelken, Halina. 1980. *Alexander von Humboldt: His Portraits and Their Artists, A Documentary Iconography*. Berlin: Dietrich Reimer Verlag.

Noda, Kesa. 1981. *Yamato Colony: 1906–1960, Livingston, California*. Livingston, Calif.: Livingston-Merced JACL Chapter.

Norbeck, Edward. 1959. *Pineapple Town: Hawaii*. Berkeley: University of California Press.

Northrup, David. 1995. *Indentured Labor in the Age of Imperialism, 1834–1922*. Cambridge: Cambridge University Press.

Nott, Josiah Clark, and George R. Gliddon. 1857. *Indigenous Races of the Earth; Or, New Chapters of Ethnological Inquiry*. Philadelphia: J. B. Lippincott.

Okihiro, Gary Y. 1991. *Cane Fires: The Anti-Japanese Movement in Hawaii, 1865–1945*. Philadelphia: Temple University Press.

———. 1994. *Margins and Mainstreams: Asians in American History and Culture.* Seattle: University of Washington Press.

———. 2001. *The Columbia Guide to Asian American History.* New York: Columbia University Press.

———. 2008. *Island World: A History of Hawai'i and the United States.* Berkeley: University of California Press.

Osborne, Thomas J. 1981. *"Empire Can Wait": American Opposition to Hawaiian Annexation, 1893–1898.* Kent, Ohio: Kent State University Press.

Osorio, Jonathan Kay Kamakawiwo'ole. 1996. "A Hawaiian Nationalist Commentary on the Trial of the *Mo'iwahine.*" In *Trial of a Queen: 1895 Military Tribunal.* Honolulu: Judiciary History Center.

———. 2002. *Dismembering Lāhui: A History of the Hawaiian Nation to 1887.* Honolulu: University of Hawai'i Press.

Pagden, Anthony. 1993. *European Encounters with the New World.* New Haven, Conn.: Yale University Press.

Paludan, Lis. 1995. *Crochet: History & Technique.* Loveland, Colo.: Interweave Press.

Pearson, Charles H. 1893. *National Life and Character.* London.

Petersen, James B. 1997. "Taino, Island Carib, and Prehistoric Amerindian Economies in the West Indies: Adaptations to Island Environments." In *The Indigenous Peoples of the Caribbean,* edited by Samuel M. Wilson, 118–30. Gainesville: University Press of Florida.

Petri, William A., Jr. 2004. "America in the World: 100 Years of Tropical Medicine and Hygiene." *American Journal of Tropical Medicine and Hygiene* 71, no. 1: 2–16.

Pickersgill, Barbara, and Charles B. Heiser Jr. 1977. "Origins and Distribution of Plants Domesticated in the New World Tropics." In *Origins of Agriculture,* edited by Charles A. Reed, 803–35. The Hague, Netherlands: Mouton.

[Piilani]. 2001. *The True Story of Kaluaikoolau as Told by His Wife, Piilani,* translated by Frances N. Frazier. Lihu'i: Kaua'i Historical Society.

Pimentel, Juan. 2000. "The Iberian Vision: Science and Empire in the Framework of a Universal Monarchy, 1500–1800." In *Nature and Empire: Science and the Colonial Enterprise,* edited by Roy MacLeod, 17–30. Special Issue of *Osiris,* 2d series, no. 15.

"The Pineapple." *Dwights American Magazine,* June 27, 1846, 328.

"The Pineapple." *Family Magazine* 4 (1836): 396.

"Pineapple Doily." *Ohio Farmer,* April 13, 1899, 331.

"Pineapple Insertion." *Ladies' Home Journal,* July 1886, 4.

"Pineapple Lace." *Ohio Farmer,* May 29, 1886, 366.

"Pineapples." *Independent,* December 13, 1877, 30.

"Pineapples in Paradise." *Fortune* 11, no. 5 (November 1930): 33–36.

Pope, Daniel. 1983. *The Making of Modern Advertising.* New York: Basic Books.

Potter, Annie Louise. 1990. *A Living Mystery: The International Art & History of Crochet.* A.J. Publishing International.

Powell, J.H. 1949. *Bring Out Your Dead: The Great Plague of Yellow Fever in Philadelphia in 1793.* Philadelphia: University of Pennsylvania Press.

Prashad, Vijay. 2001. *Everybody Was Kung Fu Fighting: Afro-Asian Connections and the Myth of Cultural Purity.* Boston: Beacon Press.

Prest, John. 1981. *The Garden of Eden: The Botanic Garden and the Re-Creation of Paradise.* New Haven, Conn.: Yale University Press.

Rhoads, William Bertolet. 1974. "The Colonial Revival." Ph.D. dissertation, Princeton University.

Rogozinski, Jan. 1999. *A Brief History of the Caribbean: From the Arawak and the Carib to the Present.* New York: Facts On File.

"Romance in the Canning Industry." *Current Opinion,* November 1, 1924, 642–43.

Root, Waverley, and Richard de Rochemont. 1976. *Eating in America: A History.* New York: William Morrow.

"Running the Charleston Blockade." *New York Observer and Chronicle,* March 2, 1865, 69.

Russ, William Adam, Jr. 1959. *The Hawaiian Revolution (1893–94).* Selinsgrove, Pa.: Susquehanna University Press.

Saville, Jennifer. 1990. *Georgia O'Keeffe: Paintings of Hawai'i.* Honolulu: Honolulu Academy of Arts.

Scanlon, Jennifer. 1995. *Inarticulate Longings: The* Ladies' Home Journal, *Gender, and the Promises of Consumer Culture.* New York: Routledge.

Schiebinger, Londa. 2004. *Plants and Empire: Colonial Bioprospecting in the Atlantic World.* Cambridge, Mass.: Harvard University Press.

Schmidt, Benjamin. 2001. *Innocence Abroad: The Dutch Imagination and the New World, 1570–1670.* Cambridge: Cambridge University Press.

Schmitt, Robert C. 1968. *Demographic Statistics of Hawaii: 1778–1965.* Honolulu: University of Hawai'i Press.

Selden, Joseph J. 1963. *The Golden Fleece: Selling the Good Life to Americans.* New York: Macmillan.

Semple, Ellen Churchill. 1911. *Influences of Geographic Environment: On the Basis of Ratzel's System of Anthropo-Geography.* New York: Holt, Rinehart and Winston.

Shah, Nayan. 2001. *Contagious Divides: Epidemics and Race in San Francisco's Chinatown.* Berkeley: University of California Press.

Shriver, J. Alexis. 1915. *Pineapple-Canning Industry of the World.* U.S. Department of

Commerce, Special Agents Series No. 91. Washington, D.C.: Government Printing Office.

Silva, Noenoe K. 2004. *Aloha Betrayed: Native Hawaiian Resistance to American Colonialism*. Durham, N.C.: Duke University Press.

Smith, Neil, and Anne Godlewska. 1994. "Introduction: Critical Histories of Geography." In *Geography and Empire,* edited by Anne Godlewska and Neil Smith, 1–8. Oxford, England: Blackwell.

Stack, George N. 1865. "Culture of the Pineapple." *Horticulturist and Journal of Rural Art and Rural Taste* 20 (May): 151–52.

Stepan, Nancy Leys. 2001. *Picturing Tropical Nature*. Ithaca, N.Y.: Cornell University Press.

Stephan, John J. 1984. *Hawaii under the Rising Sun: Japan's Plans for Conquest after Pearl Harbor.* Honolulu: University of Hawai'i Press.

Stillman, Amy K. 1989. "History Reinterpreted in Song: The Case of the Hawaiian Counterrevolution." *Hawaiian Journal of History* 23: 1–30.

Stoddard, T. Lothrop. 1914. *The French Revolution in San Domingo*. Boston: Houghton Mifflin.

———. 1920. *The Rising Tide of Color against White World-Supremacy*. New York: Charles Scribner's Sons.

Stoneman, Richard. 1991. "Introduction." In *The Greek Alexander Romance,* translated by Richard Stoneman, 1–32. London: Penguin Books.

"Study of Tropical Diseases." *Baptist Missionary Magazine* 87, no. 3 (March 1907): 101.

Takaki, Ronald. 1983. *Pau Hana: Plantation Life and Labor in Hawaii, 1835–1920.* Honolulu: University of Hawai'i Press.

Tanaka, Stefan. 1993. *Japan's Orient: Rendering Pasts into History*. Berkeley: University of California Press.

TenBruggencate, Jan K. 2004. *Hawai'i's Pineapple Century: A History of the Crowned Fruit in the Hawaiian Islands.* Honolulu: Mutual.

Thompkins, E. Berkeley. 1970. *Anti-Imperialism in the United States: The Great Debate, 1890–1920.* Philadelphia: University of Pennsylvania Press.

Thomson, J. Oliver. 1948. *History of Ancient Geography*. Cambridge: Cambridge University Press.

Thoreau, Henry David. 1906. *The Journal of Henry D. Thoreau.* Vol. 14, edited by Bradford Torrey and Francis H. Allen. Boston: Houghton Mifflin.

Thurston, Lorrin A. 1893. "The Sandwich Islands." *North American Review* 156, no. 436 (March): 265–81.

Tinker, Hugh. 1974. *A New System of Slavery: The Export of Indian Labour Overseas, 1830–1920.* London: Oxford University Press.

Tozer, H. F. 1935. *A History of Ancient Geography,* 2nd ed. London: Cambridge University Press.

Trexler, Richard C. 1995. *Sex and Conquest: Gendered Violence, Political Order, and the European Conquest of the Americas.* Ithaca, N.Y.: Cornell University Press.

"Tropical Fruit." *The Friend,* August 11, 1877, 411–12.

"Tropical Fruits." *Littell's Living Age,* March 15, 1884, 700.

Twain, Mark. [February 1901] 2002. "To the Person Sitting in Darkness." In *Vestiges of War: The Philippine-American War and the Aftermath of an Imperial Dream, 1899–1999,* edited by Angel Velasco Shaw and Luis H. Francia, 57–68. New York: New York University Press.

U.S. Department of Labor. 1931. *Labor Conditions in the Territory of Hawaii, 1929–1930.* Bureau of Labor Statistics. Bulletin No. 534. Washington, D.C.: Government Printing Office.

Walker, Nancy A. 2000. *Shaping Our Mothers' World: American Women's Magazines.* Jackson: University Press of Mississippi.

Wallerstein, Immanuel. 1989. *The Modern World-System.* Vol. 3, *The Second Era of Great Expansion of the Capitalist World-Economy, 1730–1840s.* New York: Academic Press.

Walsh, George E. 1889. "Pineapple Cultivation." *Independent,* January 31, 30.

Ward, Robert De Courcy. 1899. "Notes on Climatology." *Journal of the American Geographical Society of New York* 31, no. 1: 160–62.

———. 1908. *Climate, Considered Especially in Relation to Man.* New York: G. P. Putnam's Sons.

[Warder, Ann]. 1893. "Extracts from the Diary of Mrs. Ann Warder." *Pennsylvania Magazine of History and Biography* 17, no. 4: 444–62.

Watters, David R. 1999. "Maritime Trade in the Prehistoric Eastern Caribbean." In *The Indigenous Peoples of the Caribbean,* edited by Samuel M. Wilson, 88–99. Gainesville: University Press of Florida.

Welch, Richard E., Jr. 1979. *Response to Imperialism: The United States and the Philippine-American War, 1899–1902.* Chapel Hill: University of North Carolina Press.

"The West India Fruit Trade." *Ohio Farmer,* July 18, 1868, 454–55.

White, Henry A. 1957. *James D. Dole: Industrial Pioneer of the Pacific, Founder of Hawaii's Pineapple Industry.* New York: Newcomen Society in North America.

Williams, Mary E., and Katharine Rolston Fisher. 1911. *Elements of the Theory and Practice of Cookery.* New York: Macmillan.

Wilson, David J. 1999. *Indigenous South Americans of the Past and Present: An Ecological Perspective.* Boulder, Colo.: Westview Press.

Withey, Lynne. 1987. *Voyages of Discovery: Captain Cook and the Exploration of the Pacific.* Berkeley: University of California Press.

Woodruff, Chas. E. 1905. *The Effects of Tropical Light on White Men.* New York: Rebman.

Worcester, Dean C. 1914. *The Philippines Past and Present.* Vol. 2. New York: Macmillan.

Yun, Lisa. 2008. *The Coolie Speaks: Chinese Indentured Laborers and African Slaves in Cuba.* Philadelphia: Temple University Press.

Zalburg, Sanford. 1979. *A Spark Is Struck! Jack Hall and the ILWU in Hawaii.* Honolulu: University of Hawai'i Press.

Zamora, Margarita. 1990–91. "Abreast of Columbus: Gender and Discovery." *Cultural Critique* 17 (Winter): 127–49.

Zuckerman, Mary Ellen. 1998. *A History of Popular Women's Magazines in the United States, 1792–1995.* Westport, Conn.: Greenwood Press.

INDEX

Addams, Jane, 59

advertising: art, 148–49, 159–60, 161*fig,*
162*fig,* 209n46; genderized, 156–59, 160–
63, 180; *Ladies' Home Journal,* 144–49,
145*fig,* 146*fig,* 147*fig,* 157–58, 163, 211–
12nn20,21; pineapple, 143–50, 145*fig,*
146*fig,* 147*fig,* 153, 163, 175–76, 216n71;
with recipes, 144–48, 158–59, 165, 209n40

Africa: climate-based determinism, 1, 10, 11–
13; nationalisms, 69; plant migrations, 73,
92, 111, 112; slave labor, 188n19; sleeping
sickness, 53; "world island," 1

African Americans: Civil War troops, 18–19,
124; Du Bois, 68, 70*fig,* 118–19; Free African Society, 51; Hampton Institute, 18–19,
184n46; and Japan as champion of world's
colored peoples, 68–69, 118–19; migration
from South to the North, 172; and Philadelphia "great plague," 51; Philippines, 68; slaves,
38, 50, 62, 71, 174; "white man's burden," 58*fig*

agriculture: agricultural science, 176; America's
migrating food crops, 73; in boys' education, 107; Caribbean crop varieties, 45, 49,
80; *conuco,* 80; foreigners in Hawai'i, 106;
Hawaiian, 95–96, 176–77; Hawaiian products, 114–15; Royal Hawaiian Agricultural
Society, 114, 176; "white man's burden," 178.
See also pineapple culture; plantations; plant
management; tropical products

Ah Lim, 132

ahupua'a, 94*map,* 95–96

Alexander & Baldwin, 99–100

Alexander the Great, 2, 5, 8–9, 21, 111, 173

Alexandria: library, 9; Roman trade, 111

Allen, Richard, 51

Aloha 'Oe (Farewell to Thee), Lili'uokalani's,
127

"aloha spirit," 176

American Anti-Imperialist League, 59

American Can Company, 133, 154

American Cookery (Simmons), 158

American Eugenics Society, 18

American Factors, 100

American Federation of Labor, 59

American Gardener's Magazine, 167

American Indians: Aruak people, 78; Callinago,
80, 174; climate-based determinism, 10, 11,
13, 34–35; Columbus's imaginary, 2, 27, 28–
29, 174; European views of, 13, 27, 28–38,
35*fig;* Hampton Institute, 184n46; as hunter-
gatherers, 74–78; as inferior, 13, 31, 33;
Ka'apor people, 77–78; noble savage, 29–30,
35–36; pineapple, 74–81, 79*map;* sea otter
pelts, 100; slaves, 29, 31, 188n19; Tupí-
Guaraní, 74–78, 174, 215n65; Tupinamba,
75*fig;* "white man's burden," 58*fig*

American Medical Association, 63

American Museum of Natural History, 60

Designer:	Sandy Drooker
Cartographer:	Bill Nelson
Indexer:	Barbara Roos
Text:	10/14 Adobe Garamond
Display:	Bank Gothic Medium, Caslon Antiqua
Compositor:	Integrated Composition Systems
Printer and binder:	Maple-Vail Manufacturing Group